AF488034

ANALOG ELECTRONIC PRINCIPLES

FIRST EDITION

Author

Dr. Amit Tomar (IIT Alumnus)
M.Sc, M.Phil, Ph.D, NET-JRF, GATE
Senior Assistant Professor
Department of Physics and Astronomical Sciences
Central University of Jammu (Government of India)

Co Author

Dr. Sandeep Arya
M.Sc., Ph.D., NET
Senior Assistant Professor
Department of Physics
University of Jammu

INDIA • SINGAPORE • MALAYSIA

ISBN 979-8-88684-650-8

Contents

Preface

We are delighted to announce the publication of the first edition of our book "Analog Electronics."

We are currently teaching at two different universities, and this book is primarily a compilation of our classroom teachings during the past 15 years. While framing the book suggestions from both undergraduate and postgraduate students were incorporated.

The goal of this book is to explain the fundamental concepts of analog electronics to Undergraduate, B.Sc Honors, and Postgraduate students at both University and college levels.

It can be used as a textbook or as a supplement to any comparable text currently available.

Anyone interested in the subject should find it useful for self-study. This book is very useful for B.Sc, B.Sc Honors(Physics), and M.Sc Physics Students. It presents the textual explanations and concepts from the textbook in a clear, simple format.

The book is divided into various chapters, the subject matter is simple to grasp, and it is designed with a special emphasis on newcomers at the graduate level. Each chapter goes into great detail about each and every idea. Wherever necessary, circuit diagrams are supplied in a clear and precise manner.

The book's language is logical and straightforward, allowing even a novice to follow along.

A number of authors and publishers whose works were freely consulted throughout the production of this book owe us a duty of gratitude.

We will be thankful to readers for pointing up any inaccuracies or omissions that may have sneaked in despite our best efforts.

I would like to thank the co-author of my book, **Dr. Sandeep Arya**, Sr. Assistant Professor, Department of Physics, University of Jammu, for his Invaluable contributions and suggestions throughout the book.

Suggestions for improving the book are gratefully accepted.

– Dr. Amit Tomar
Sr. Assistant Professor
Department of Physics and Astronomical Sciences
Central University of Jammu
Govt. of India.

Chapter 1

Diode Circuit Theory

A Brief Chapter Overview:

1.1	Introduction to Diode
1.2	Clipper Circuit
1.3	Clamping Circuit
1.4	Rectifier
1.5	Zener Diode as Voltage Regulator

1.1 Introduction to Diode

A diode is an equipment which allows the electric current to pass in one direction only. In the other direction it produces a very high resistance thereby hindering the current to flow in reverse direction. Cathode and Anode, the two electrodes, constitute the diode thereby giving it an identical name.

Diodes have many modern applications and are frequently used for radio conversion, logic gates, rectification, temperature measurements and current steering etc.

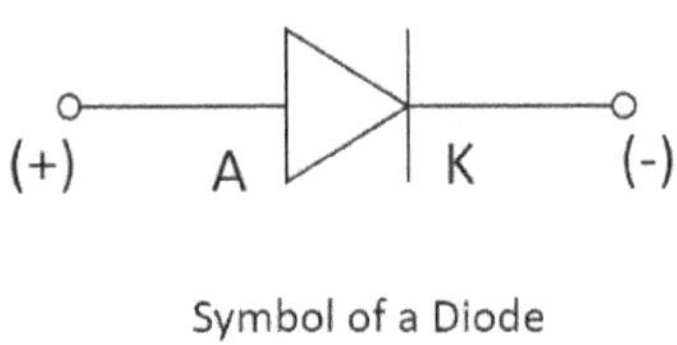

Fig. 1.1: Symbol for a typical diode

Anode, the positive terminal of the diode is represented by A whereas Cathode, which is the negative terminal is represented by K in the above figure.

1.2 Formation of a Diode

A diode is basically two differently doped semiconductor material namely P-type and N-type sandwiched together. When these materials are brought together, they recombine together to form a junction, as shown in the figure;

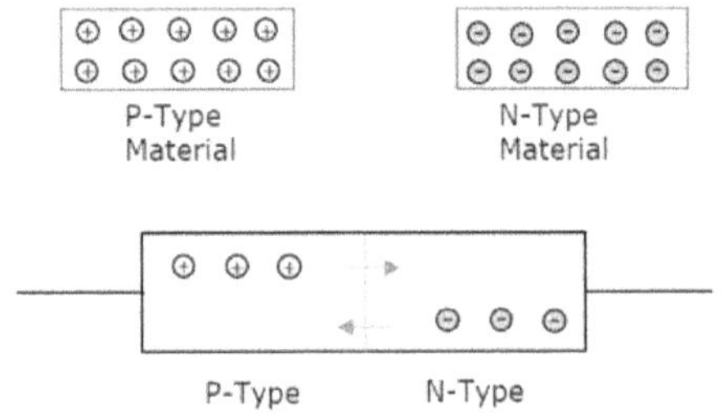

Fig.1.2: P-type and N-type material

In a P-type material, the majority charge carriers are the holes whereas in an N-type material, the majority charge carriers are the electrons.

As the opposite charges attracted towards each other, some of the holes from P-type material tends to move to N-type region. Similarly, some electrons from N-type material tend to move to P-type region. As both these charges approach the junction, they recombine with each other on either side of the junction to form ions. These positive and negative ions formed on either side of the junction are immovable in nature thereby resulting in formation of the depletion region. Depletion region is the region of immovable charges. These ions are stationary and maintain a region of space between them without any charge carriers.

Due to the different charge distribution on both side of the junction, there arose a potential difference (V). This potential acts as a barrier as it further prevents the movement of electrons and holes across the junction. This is called the potential barrier of the junction.

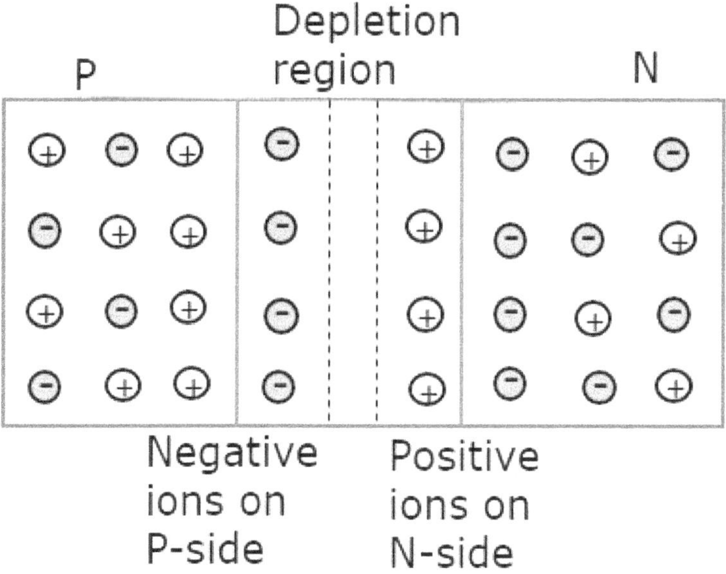

Fig.1.3: Depletion layer in a PN junction

1.3 Biasing of a Diode

There are two possible conditions in which a diode can be biased;

- Forward biased

- Reverse biased

Forward Bias Condition

When the P-type region (anode) of the diode is connected to the positive terminal and the N-Type region (cathode) is connected to the negative terminal of the voltage source, then the diode is said to be forward biased. A diode conducts effectively in forward biased condition.

Reverse Bias Condition

When the P-type region of the diode is connected to the negative terminal and the N-type region is connected to the positive terminal of the voltage source, then the diode is said to be reverse biased. A diode merely conducts in reverse biased condition.

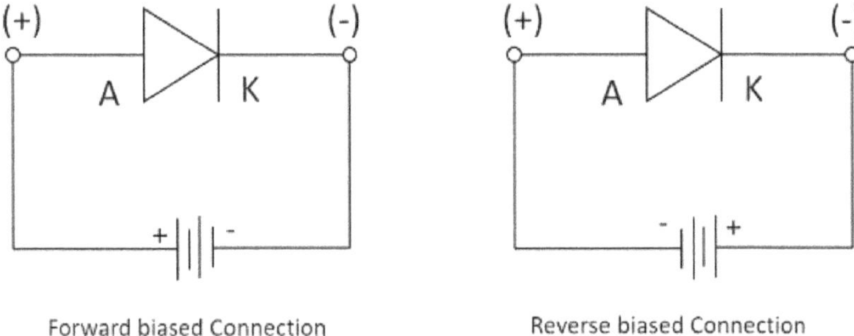

Fig.1.4: A typical diode in forward and reverse biased condition

The working of a diode depends upon its condition of biasing. Let's try to find out how a typical diode works under different conditions of biasing.

Working under Forward Biased

Under forward biasing, the holes in P-type and electrons in N-type tend to move across the junction, ultimately collapsing the potential barrier. When the barrier is collapsed, there starts a free flow of current across the junction.

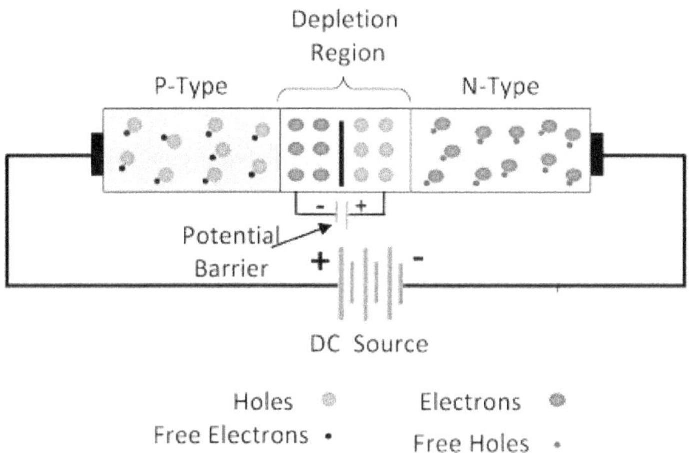

Fig: 1.5 Forward biasing

But in order to break the potential barrier, the source voltage must be equivalent or more than the barrier. Once, the barrier is broken, current is constituted and is termed as forward current.

Forward Current is the current generated by diode operating under forward biasing. It is generally denoted by I_f.

Working under Reverse Biased

Under reverse bias condition, the depletion layer is thickened and the flow of charges is restricted. This is because, when anode and cathode are connected to negative and positive ends of source respectively, the electrons are attracted towards positive end and holes are attracted towards negative end. Hence, both of them will move away from the barrier thereby increasing the junction resistance and preventing any electron to cross the junction.

Following figure explains this behaviour.

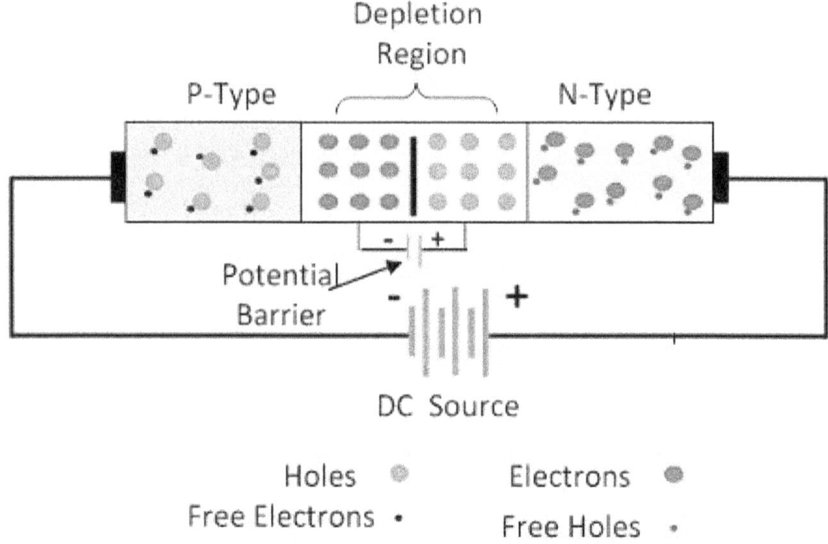

Fig: 1.6 Reversed biasing

The number of minority carriers crossing the junction decreases with the corresponding increase in the reverse potential. This also constitutes a current termed as reverse current. This current is generally negligible and is almost constant under controlled temperature. Reverse current is the current generated by a diode operating under reverse biasing. It is generally denoted by I_r.

When we keep on increasing the reverse voltage, then at a point a sudden breakdown of the diode occurs accompanied with a large amount of current across the junction. This is called the reverse breakdown of the diode. It happen under two different phenomenon:

- Zener breakdown

- Avalanche breakdown

This high reverse current can damage the diode. The maximum reverse voltage that a diode can withstand without being damaged is known as peak inverse voltage.

It is an important parameter for the reverse breakdown condition. Hence, it is considered during reverse biased condition. It signifies how a diode can be safely operated in reverse bias.

From above discussion, it is clearly understood that a diode acts as a conductor during forward biasing and acts as an insulator during reverse biasing. This typical nature of a diode makes it useful for many modern day applications.

1.4 V - I Characteristics of a Diode

A typical circuit arrangement for a PN junction diode is depicted by the following figure. Here, ammeter is connected in series and voltmeter in parallel, while the supply is controlled through a variable resistor.

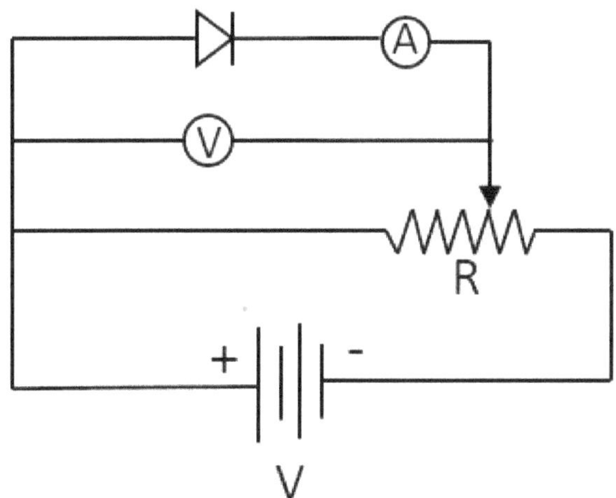

A Practical diode circuit

Fig:1.7 A typical diode circuit

When a PN junction diode is operated under forward biased condition, at some particular value of voltage, the potential barrier across the junction gets disappeared. That voltage is called Knee Voltage or cut-off voltage.

The value of the knee voltage depends upon the nature of the material. For example; the value of knee voltage for silicon is 0.7V whereas for germanium is 0.3V. This is the reason why silicon is preferred over germanium in modern electronics.

When the forward voltage exceeds beyond the knee voltage, the forward current rises up exponentially and if this is done further, the equipment can get damaged due to overheating.

Similarly, when the diode is operated under reverse biased condition, very tiny current produced by the minority carriers exist. This is known as reverse current. As the reverse voltage increases, the reverse current increases and suddenly breaks down at a point, resulting in the permanent elimination of the junction.

The behaviour of a diode can be easily understood by its V-I characteristics. It is plot between the input voltage and the output current. Following figure shows the V-I characteristics of a PN junction diode under both biased conditions;

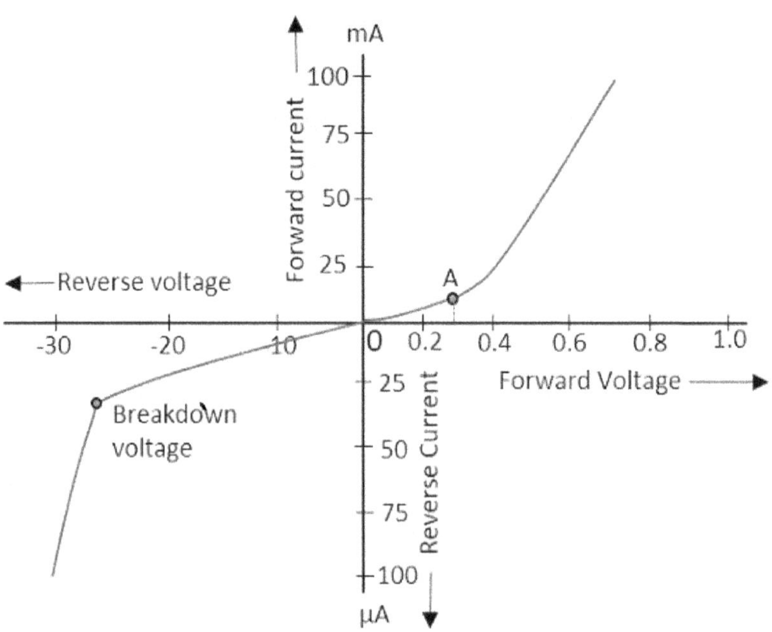

Fig: 1.8 V-I characteristics of a PN junction

Ideal diode

An ideal diode is a two-terminal electronic component that conducts current primarily in only one direction i.e., it has very low (ideally zero) resistance in one direction, and very large (ideally infinite) resistance in the other.

An ideal diode is a hypothetical concept but it helps in attaining more efficiency in modern electronics. Semiconductor diodes were the first semiconductor electronic device. It was Ferdinand Braun, a German physicist, who first discovered asymmetric electrical conduction across the contact between a crystalline mineral and a metal. Today, most diodes are made of silicon, but other materials such as gallium arsenide and germanium are used.

1.5 Clipper Circuit

A clipper is a device that removes or limits some portion of the input wave form (input voltage signal) above or below a certain given level. In other words, a circuit which limits positive or negative amplitude or both from the input signal is called clipper circuit.

There are four kinds of clipper circuit;

- Positive clipper circuit

- Negative clipper circuit

- Biased clipper circuit

- Positive and negative clipper circuit

• Positive Clipper Circuit

A Clipper circuit which is used to limit positive portion of the given input signal is known as positive clipper circuit. Following figures depicts a positive clipper circuit and its input and output waveforms;

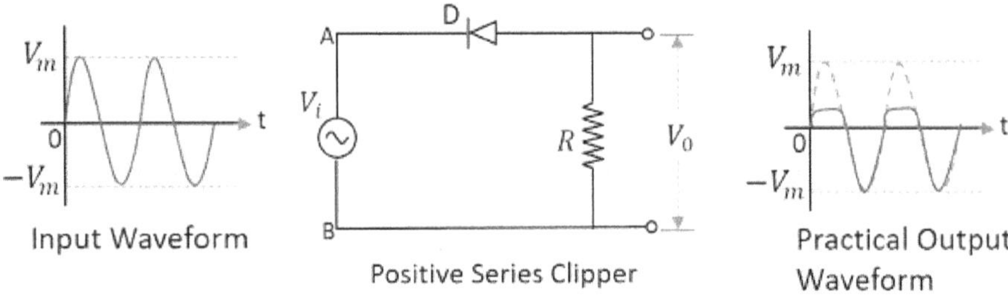

Input Waveform

Positive Series Clipper

Practical Output Waveform

From the above waveforms, it can be seen that only a part of the positive peak was clipped off. This is because of the voltage across V_0. But for the ideal case, the output was not meant to be like this. Following figures point out the difference between practical and ideal output waveforms.

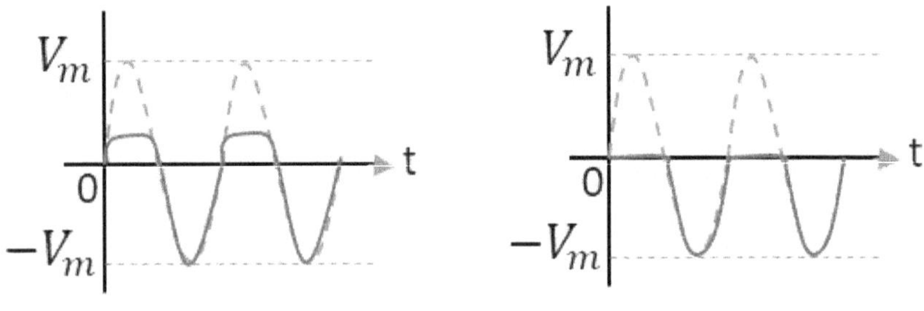

Practical Output Waveform

Ideal Output Waveform

Here, unlike in the case of the ideal diode, a bit portion of the positive part of cycle is present in the practical output. This is due to the fact that the knee voltage for a practical diode (for e.g., silicon diode) is 0.7v whereas the knee voltage for ideal diode is 0V. Hence, the difference can be in both the waveforms.

• Negative clipper circuit

A Clipper circuit which is used to limit only the negative part of given input signal is known as a negative clipper circuit. Following figures represents a negative clipper circuit and its input and output waveforms;

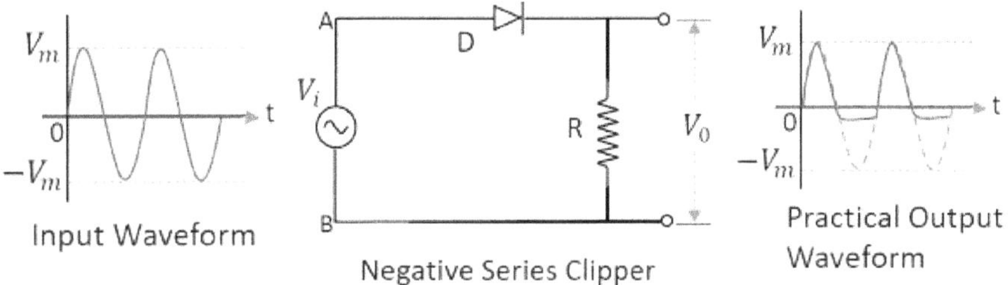

Input Waveform

Negative Series Clipper

Practical Output Waveform

From the above waveforms, it is clear that only a part of the negative output signal is clipped, not the entire negative part. This is somewhat different from the ideal case. Following figures point out the difference between the practical and ideal output waveforms;

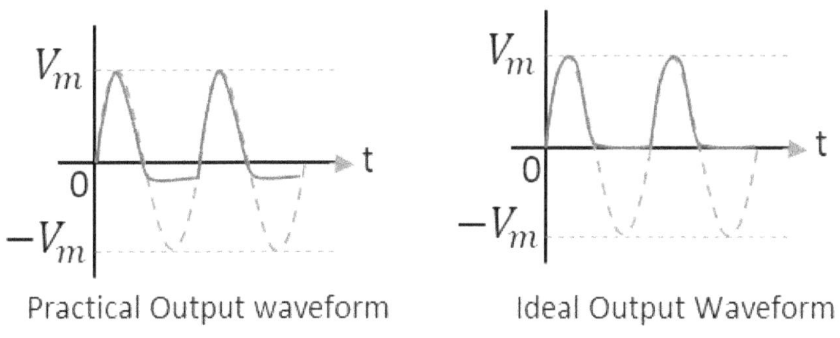

Practical Output waveform

Ideal Output Waveform

Here, unlike the ideal output waveform where whole negative output is clipper, a bit portion of the negative part of the cycle is still remaining in the practical output waveform. This is again due to the fact that a practical diode operates at certain value of its knee voltage ($\approx 0.7V$) which is not there in the ideal case. Hence there is a divergence in the practical and ideal output waveforms.

- **Biased clipper circuit**

Sometimes, clipping is required for different biasing levels. So in order to-do this, a biased clipper circuit is required. For biased clipping circuits for voltage waveforms to operate at different levels, a bias voltage (V_{BIAS}) is given in series with the diode to produce a combination clipper called the biased clipper.

The voltage across the series combination must be greater than the sum of V_{BIAS} and the knee voltage of the diode ($\approx 0.7V$) before the diode becomes sufficiently forward biased to conduct.

For e.g., if V_{BIAS} is fixed at 5.0V, then the total voltage at the positive terminal must be greater than 5.0+ 0.7 i.e., 5.7 volts for it to become forward biased. Any voltage levels above this bias point are clipped off.

There are few kinds of biased clipping;

Positive Bias Diode Clipping

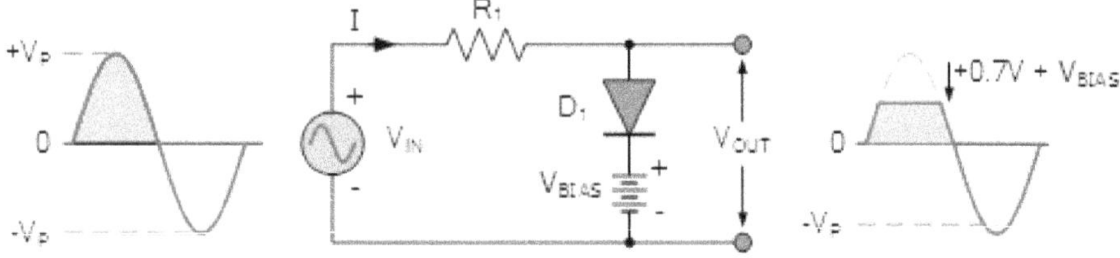

Negative Bias Diode Clipping

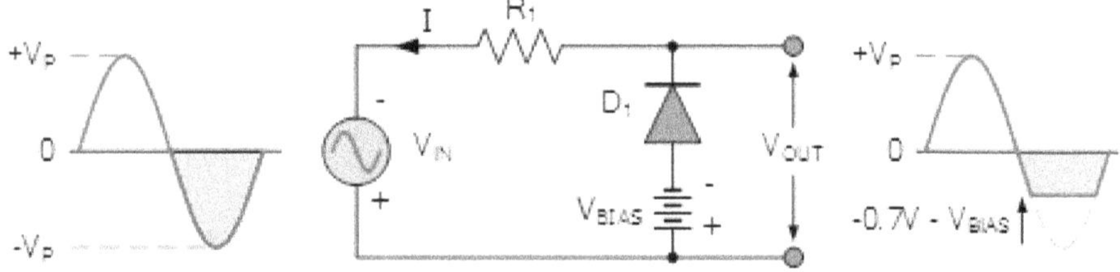

• Combination clipper circuit

A variable diode clipping can be obtained by changing the bias voltage of the diodes. If both the positive and the negative half cycles are to be clipped, then the combination of two biased clipping diodes are used. This results in a combination clipper circuit. When a portion of both positive and negative of each half cycle of the input voltage is to be clipped (or removed), combination clipper is employed. Circuit diagram for combination clipper is shown in figure.

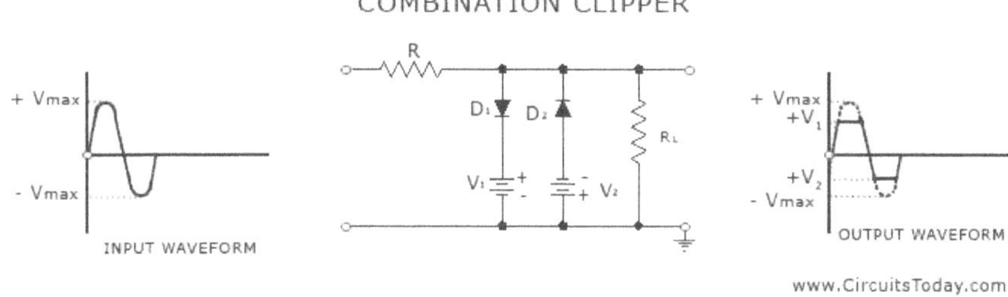

Applications of clipper circuits

Clipper circuits are commonly used;

- For generation and shaping of waveforms

- For the protection of circuits from spikes

- For amplitude restorers

- In FM transmitters

- In television circuits

- As voltage limiters

1.6 Clamping Circuit

A Clamper circuit is a combinational circuit of a diode, a resistor and a capacitor that transforms the input waveform to a desired DC level without altering the applied signal. It actually additions a DC level to an AC signal. Here, the positive and negative peaks of the signals can be arranged at desired levels using the clamping circuits. As the DC level gets shifted, it is also termed as a Level Shifter.

Types of Clamping circuits;

These are two common types of clamping circuits

- **Positive Clamper**

   Positive clamper with $+V_r$

   Positive clamper with $-V_r$

- **Negative Clamper**

  Negative clamper with $+V_r$

  Negative clamper with $-V_r$

• Positive Clamper Circuit

A Positive Clamper circuit is one that consists of a diode, a resistor and a capacitor. It shifts the output signal to the positive portion of the input signal. The figure below explains the construction of a positive clamper circuit.

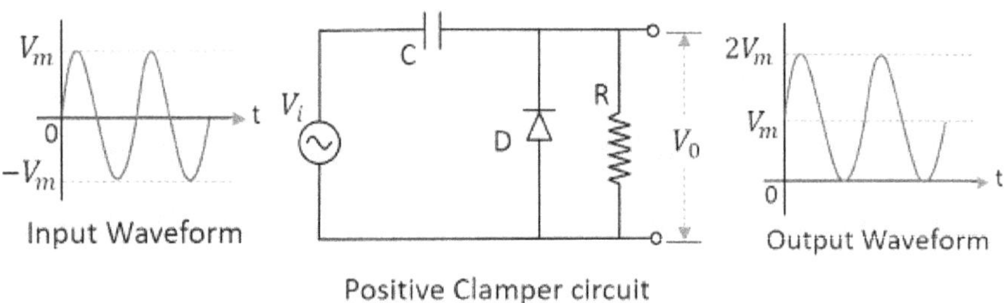

Positive Clamper circuit

Initially when the input is given, the capacitor is not yet charged and the diode is reverse biased. The output is not considered at this point of time. During the negative half cycle, at the peak value, the capacitor gets charged with negative on one plate and positive on the other. The capacitor is now charged to its peak value V_m. The diode is forward biased and conducts heavily.

During the next positive half cycle, the capacitor is charged to positive V_m while the diode gets reverse biased and gets open circuited. The output of the circuit at this moment will be

V (out) $= V_i + V_m$

Hence the signal is positively clamped as shown in the above figure. The output signal changes according to the changes in the input, but shifts the level according to the charge on the capacitor, as it adds the input voltage.

- **Positive Clamper with (+ V_r)**

A Positive clamper circuit if biased with some positive reference voltage, that voltage will be added to the output to raise the clamped level. Using this, the circuit of the positive clamper with positive reference voltage is constructed as below.

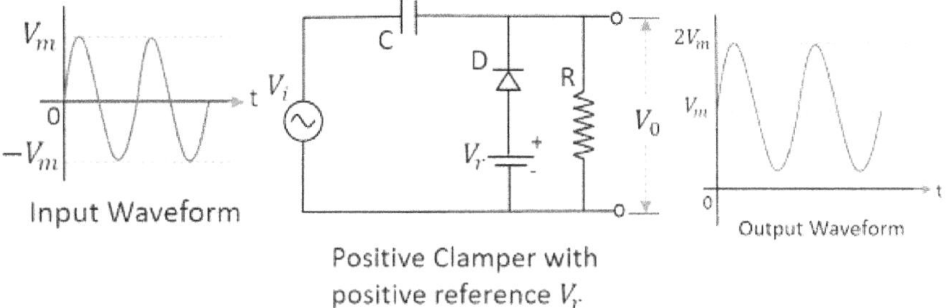

Positive Clamper with
positive reference V_r.

During the positive half cycle, the reference voltage is applied through the diode at the output and as the input voltage increases, the cathode voltage of the diode increase with respect to the anode voltage and hence it stops conducting. During the negative half cycle, the diode gets forward biased and starts conducting. The voltage across the capacitor and the reference voltage together maintain the output voltage level.

- **Positive Clamper with (- V_r)**

A Positive clamper circuit if biased with some negative reference voltage, that voltage will be added to the output to raise the clamped level. Using this, the circuit of the positive clamper with positive reference voltage is constructed as below.

During the positive half cycle, the voltage across the capacitor and the reference voltage together maintain the output voltage level. During the negative half-cycle, the diode conducts when the cathode voltage gets less than the anode voltage. These changes make the output voltage as shown in the above figure.

• **Negative Clamper**

A Negative Clamper circuit is one that consists of a diode, a resistor and a capacitor and that shifts the output signal to the negative portion of the input signal. The figure below explains the construction of a negative clamper circuit.

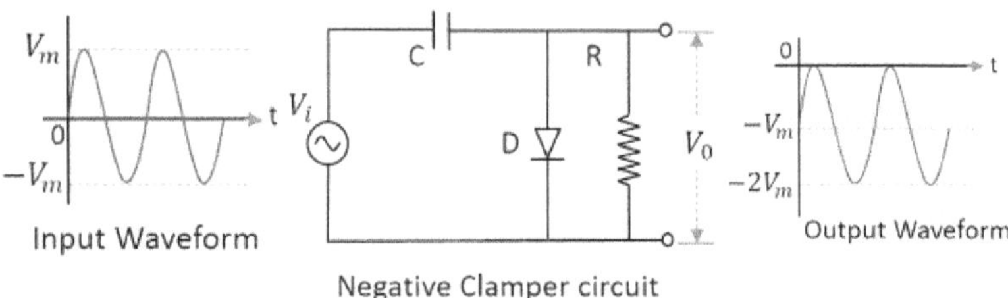

Negative Clamper circuit

During the positive half cycle, the capacitor gets charged to its peak value V_m. The diode is forward biased and conducts. During the negative half cycle, the diode gets reverse biased and gets open circuited. The output of the circuit at this moment will be

$$V(out) = V_i + V_m$$

Hence the signal is negatively clamped as shown in the above figure. The output signal changes according to the changes in the input, but shifts the level according to the charge on the capacitor, as it adds the input voltage.

Applications of clamping circuits

Clipping circuits have many modern applications. Following are some of them;

- Used as direct current restorers
- Used to remove distortions
- Used as voltage multiplier
- Used for the protection of amplifiers
- Used as test equipment
- Used as base-line stabilizer

1.7 Rectifier

A rectifier is simply electrical equipment that converts alternating current (AC) to direct current (DC). It is generally composed of one or more diodes.

Since, a diode only allows electric current to flow in one direction only, it can be used to rectify or control the output. This process is called rectification.

A rectifier can take the shape of several different physical forms such as solid-state diodes, vacuum tube diodes, mercury arc valves, silicon-controlled rectifiers and various other silicon-based semiconductor switches

Basically, there are two types of rectifiers:

- Half wave rectifier

- Full wave rectifier

Some important terms regarding a rectifier

- Peak Inverse Voltage

  We know that when a diode is operated under reverse bias, after a certain level of voltage, diode breakdown happen leading into a large inflow of current across the junction. This can damage the diode if the voltage is kept on increasing. Hence it is very important to know about that maximum voltage.

  The maximum reverse voltage that a diode can tolerate without getting damaged is known as Peak inverse voltage.

- Form Factor

  Form factor can be explained as the average of absolute values of all points on the given waveform. It is defined as the ratio of R.M.S. value to the average value of the given quantity. It is represented by F.

$$F = \frac{rms\ value}{average\ value} = \frac{I_m/2}{I_m/\pi} = \frac{0.5I_m}{0.318I_m} = 1.57$$

- Peak Factor

  Peak factor is defined as the ratio of peak value to the R.M.S. value of the given quantity.

  So,

$$PeakFactor = \frac{Peak\ value}{r.\,m.\,s\ value} = \frac{V_m}{V_m/2} = 2$$

  Peak factor is an important entity in determining the efficiency of a rectifier.

These are all important terms for studying a rectifier.

- ## Half Wave Rectifier

A half wave rectifier can only rectify half of the given input cycle, hence the name half wave rectifier. Here, input signal (AC signal) is provided with the help of an input transformer that steps up or down depending upon the need. But in most of the cases, a step down transformer is employed in rectifier circuits in order to reduce the input voltage.

The input signal given to the transformer is passed through a PN junction diode which acts as a rectifier. This diode converts the AC voltage into pulsating dc for only the positive half cycles of the input. A load resistor is connected at the end of the circuit. The figure below shows the circuit of a half wave rectifier.

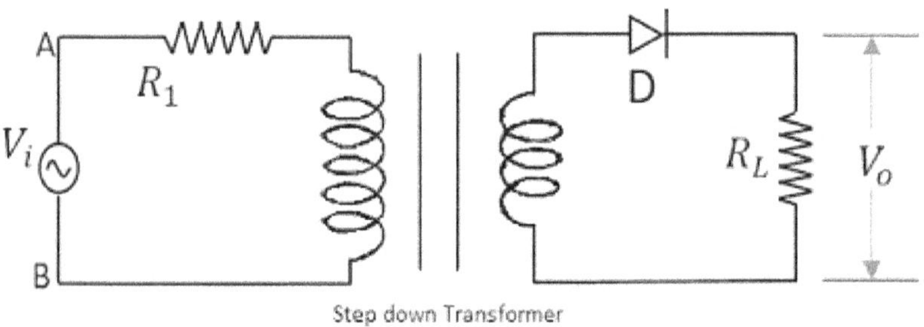

Step down Transformer

Working of a HWR

The input signal is given to the transformer which reduces the voltage levels. The output from the transformer is given to the diode which acts as a rectifier. This diode gets ON (conducts) for positive half cycles of input signal. Hence a current flows in the circuit and there will be a voltage drop across the load resistor. The diode gets OFF (doesn't conduct) for negative half cycles and hence the output for negative half cycles will be, I_D=0and V_o=0.

Hence the output is present for positive half cycles of the input voltage only (neglecting the reverse leakage current). This output will be pulsating which is taken across the load resistor.

Waveforms of a HWR

The input and output waveforms are as shown in the following figure.

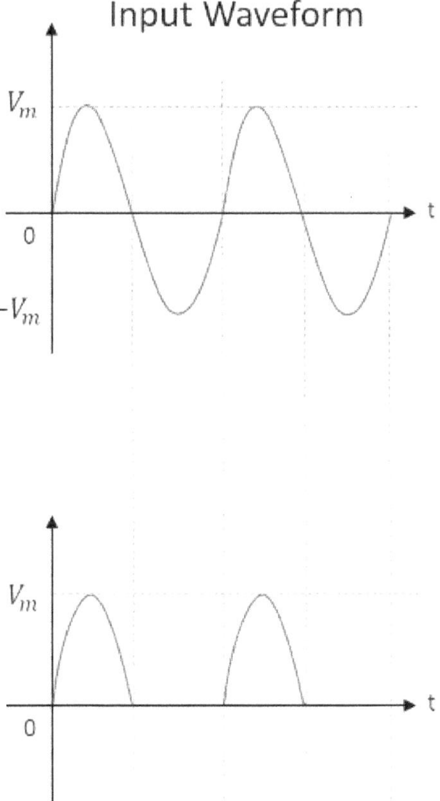

From the waveform, it can be seen that the output of a half wave rectifier is in the form of a pulsating DC. Now, let's try to figure out what is happening by analysing some observables obtained from the output

Analysis of a HW rectifier

There are some keys quantities that explain this behaviour of the rectifier.

The equation for input voltage given to the device is as follows;

$$V_i = V_m \sin \omega t$$

Where,

$$V_m \text{ is the peak value of supply voltage}$$

Let's suppose the diode to be ideal. In that case, R_f is the resistance in the ON state (forward direction) and R_r in the OFF state (reverse direction).

Now, the total current through the load resistance R_L is given by,

$$i = I_m \sin \omega t \quad for \quad 0 \le \omega t \le 2\pi$$
$$i = 0 \quad\quad for \quad \pi \le \omega t \le 2\pi$$

Where

$$I_m = \frac{V_m}{R_f + R_L}$$

Now, the DC output voltage is,

$$V_{dc} = I_{dc} \times R_L = \frac{I_m}{\pi} \times R_L$$

$$= \frac{V_m \times R_L}{\pi (R_f + R_L)} = \frac{V_m}{\pi \{1 + (R_f / R_L)\}}$$

If $R_L \gg R_f$, then

$$V_{dc} = \frac{V_m}{\pi} = 0.318 V_m$$

I_{RMS} and V_{RMS}

The root mean square value of current is given by;

$$I_{rms} = \left[\frac{1}{2\pi} \int_0^{2\pi} i^2 d(\omega t) \right]^{\frac{1}{2}}$$

$$I_{rms} = \left[\frac{1}{2\pi} \int_0^{2\pi} I_m^2 \sin^2 \omega t \, d(\omega t) + \frac{1}{2\pi} \int_\pi^{2\pi} 0 \, d(\omega t) \right]^{\frac{1}{2}}$$

$$= \left[\frac{I_m^2}{2\pi} \int_0^\pi \left(\frac{1 - \cos 2\omega t}{2} \right) d(\omega t) \right]^{\frac{1}{2}}$$

$$= \left[\frac{I_m^2}{4\pi} \left\{ (\omega t) - \frac{\sin 2\omega t}{2} \right\}_0^\pi \right]^{\frac{1}{2}}$$

$$= \left[\frac{I_m^2}{4\pi} \left\{ \pi - 0 - \frac{\sin 2\pi}{2} + \sin 0 \right\} \right]^{\frac{1}{2}}$$

$$= \left[\frac{I_m^2}{4\pi} \right]^{\frac{1}{2}} = \frac{I_m}{2}$$

$$= \frac{V_m}{2(R_f + R_L)}$$

Similarly, the root mean square value of voltage across the load resistance R_L is given by,

$$V_{rms} = I_{rms} \times R_L = \frac{V_m \times R_L}{2(R_f + R_L)}$$

$$= \frac{V_m}{2\{1 + (R_f / R_L)\}}$$

If $R_L \gg R_f$, then

$$V_{rms} = \frac{V_m}{2}$$

Rectifier Efficiency

The efficiency of any electrical circuitis the ratio of its output power to the input power.

In case of a HW rectifier, it is given by the relation;

$$\eta = \frac{d.c.\,power\ delivered\ to\ the\ load}{a.c.\,input\ power\ from\ transformer\ secondary} = \frac{P_{ac}}{P_{dc}}$$

Now

$$P_{dc} = (I_{dc})^2 \times R_L = \frac{I_m R_L}{\pi^2}$$

Further

$$P_{ac} = P_a + P_r$$

Where

$$P_a = power\ dissipated\ at\ the\ junction\ of\ diode$$

$$= I_{rms}^2 \times R_f = \frac{I_m^2}{4} \times R_f$$

And

$$P_r = power\ dissipated\ in\ the\ load\ resistance$$

$$= I_{rms}^2 \times R_L = \frac{I_m^2}{4} \times R_L$$

$$P_{ac} = \frac{I_m^2}{4} \times R_f + \frac{I_m^2}{4} \times R_L = \frac{I_m^2}{4}(R_f + R_L)$$

From both the expressions of P_{ac} and P_{dc}, we can write

$$\eta = \frac{I_m^2 R_L / \pi^2}{I_m^2 (R_f + R_L)/4} = \frac{4}{\pi^2} \frac{R_L}{(R_f + R_L)}$$

$$= \frac{4}{\pi^2} \frac{1}{\{1 + (R_f/R_L)\}} = \frac{0.406}{\{1 + (R_f/R_L)\}}$$

Percentage rectifier efficiency

$$\eta = \frac{40.6}{\{1 + (R_f/R_L)\}}$$

Theoretically, the maximum value of rectifier efficiency of a half wave rectifier is 40.6% when $R_f/R_L = 0$

Further, the efficiency may be calculated in the following way

$$\eta = \frac{P_{dc}}{P_{ac}} = \frac{(I_{dc})^2 R_L}{(I_{rms})^2 R_L} = \frac{(V_{dc}/R_L)^2 R_L}{(V_{rms}/R_L)^2 R_L} = \frac{(V_{dc})^2}{(V_{rms})^2}$$

$$= \frac{(V_m/\pi)^2}{(V_m/2)^2} = \frac{4}{\pi^2} = 0.406$$

$$= 40.6\%$$

Ripple Factor

The rectified output from the circuit still contains some amount of AC component which is there in the form of ripples. These ripples can be understood from the output waveform of HW rectifier. In order to get a pure DC output, these ripples must be considered.

The ripple factor is the measure of these ripples or waviness in the rectified output. It is denoted by **y**.

Ripple factor is given by the relation;

$$\gamma = \frac{ripple\ voltage}{d.c\ voltage} = \frac{rms\ value\ of\ a.c.\ component}{d.c.\ value\ of\ wave} = \frac{(V_r)_{rms}}{v_{dc}}$$

Here,

$$(V_r)_{rms} = \sqrt{V_{rms}^2 - V_{dc}^2}$$

Therefore,

$$\gamma = \frac{\sqrt{V_{rms}^2 - V_{dc}^2}}{V_{dc}} = \sqrt{\left(\frac{V_{rms}}{V_{dc}}\right)^2 - 1}$$

Now,

$$V_{rms} = \left[\frac{1}{2\pi} \int_0^{2\pi} V_m^2 \sin^2 \omega t\ d(\omega t)\right]^{\frac{1}{2}}$$

$$V_{dc} = V_{av} = \frac{1}{2\pi}\left[\int_0^{\pi} V_m \sin \omega t\ d(\omega t) + \int_0^{2\pi} 0.d(\omega t)\right]$$

$$= \frac{V_m}{2\pi}\left[-\cos \omega t\right]_0^{\pi} = \frac{V_m}{\pi}$$

$$\gamma = \sqrt{\left[\left\{\frac{(V_m/2)}{(V_m/\pi)}\right\}^2 - 1\right]} = \sqrt{\left\{\left(\frac{\pi}{2}\right)^2 - 1\right\}} = 1.21$$

The ripple factor is also defined as

$$\gamma = \frac{(I_r)_{rms}}{I_{dc}}$$

As the value of ripple factor for a HW rectifier is 1.21, it implies that the amount of AC component present in the output is 121% of the DC voltage.

Voltage Regulation

The current passing the load may change depending upon load resistance. But even then, we expect our output voltage, which is taken across that load resistance, to be constant. So, our voltage needs to be regulated even under different load conditions.

This variation of DC output voltage with change in DC load current is voltage regulation. It is given by the relation,

$$Percentage\ regulation = \frac{V_{no\ load} - V_{full\ load}}{V_{full\ load}} \times 100\%$$

Lower the value of percentage regulation better would be the power supply. For an ideal power supply, voltage regulation is zero.

Full Wave Rectifier

A full wave rectifier can rectify both the positive and the negative part of the input cycle. It is more commonly used type of rectifier. It can be of following forms;

- Center-tapped full wave rectifier

- Bridge full wave rectifier

Both of these are readily used depending upon their pros and cons. Let's now look into both of them for further details;

Centertapped Full-Wave Rectifier

A rectifier where secondary transformer is tapped to get the desired output voltage, using two diodes alternatively, to rectify the complete cycle is known as a Center tapped Full wave rectifier.

Some key specifications of a center tapping transformer;

1. The tapping is done by drawing a lead at the mid-point on the secondary winding. By doing so, the winding is split into two equal halves

2. The voltage at the tapped mid-point is zero. This is called the neutral point.

3. The center tapping provides two separate output voltages which are equal in magnitude but opposite in polarity to each other.

4. A number of tapings can be drawn out to obtain different levels of voltages.

Construction

Two rectifier diodes with a center tapped transformer used for the construction of a Center tapped full wave rectifier. The circuit diagram of a center tapped full wave rectifier is as shown below.

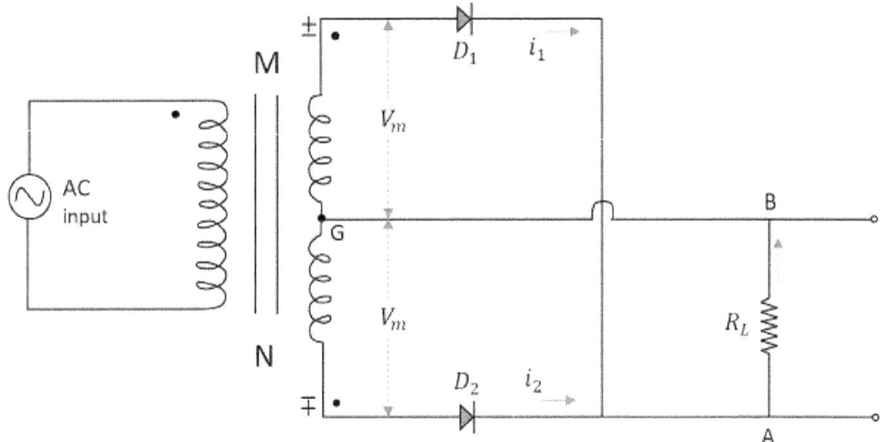

Circuit diagram of a center-tapped full wave rectifier

Fig: 1.9 Circuit diagram of a center tapped FW rectifier

Working

Following figures explains the working of the rectifier.

When the positive half cycle of the input voltage is applied, the point M at the transformer secondary becomes positive with respect to the point N. This makes the diode D1 forward biased. Hence current i_1

flows through the load resistor R_L from A to B. This gives the positive half cycles in the output

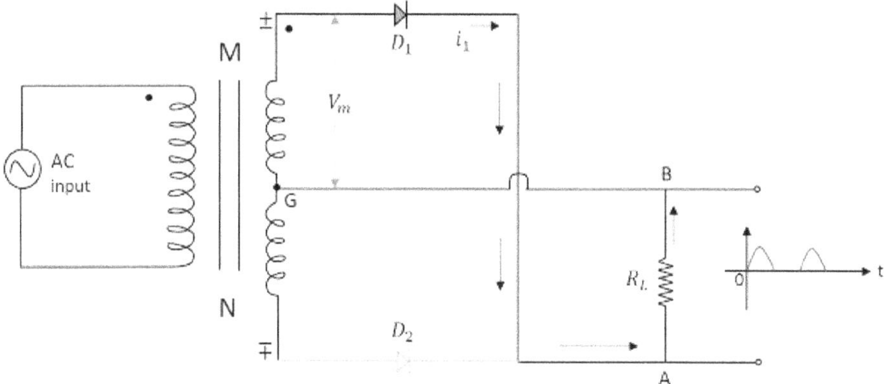

Fig: 1.10 Working for a positive half cycle of input

Similarly, when the negative half cycle of the input voltage is applied, the point M at the transformer secondary becomes negative with respect to the point N. This makes the diode D2 forward biased. Hence, current i_2 flows through the load resistor R_L from A to B. This gives the positive half cycles in the output, even during the negative half cycles of the input.

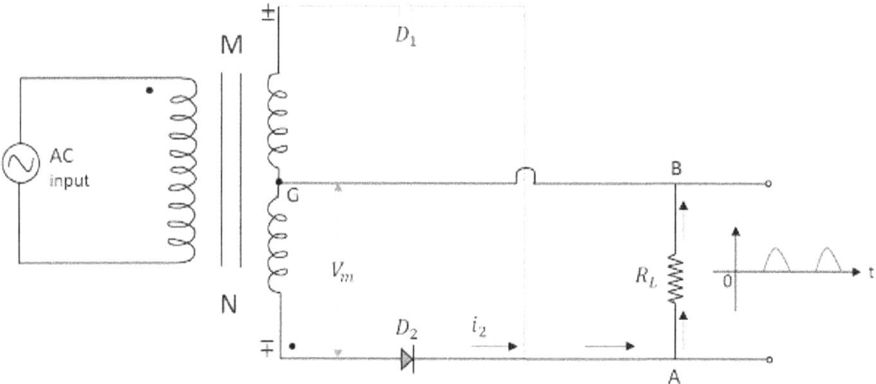

Fig: 1.11 Working for a negative half cycle of input

Waveforms

The input and output waveforms of the centertapped full wave rectifier are given below;

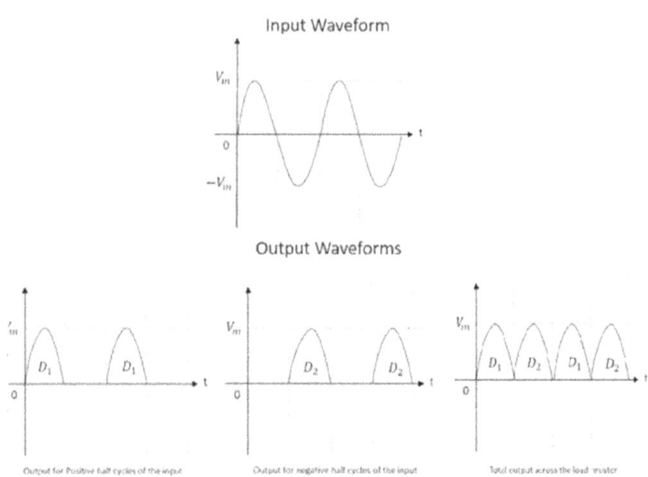

From the above figure, it is clear that the output is obtained for both the positive and negative half cycles. It is also observed that the output across the load resistor is in the same direction for both the halves of the input cycle.

PIV

As the maximum voltage across half secondary winding is V_m, the whole of the secondary voltage appears across the non-conducting diode. Hence the peak inverse voltage is twice the maximum voltage across the half-secondary winding, i.e.

$$PIV = 2V_m$$

Disadvantages of a center tapped FW rectifier

There are some disadvantages of a center-tapped full wave rectifier;

- PIV of the diodes should be high

- Location of center-tapping is difficult

- The dc output voltage is small

Bridge FullWave Rectifier

A bridge full wave rectifier uses four diodes connected in bridge form so as not only to produce the output during the full cycle of input, but also to eliminate the disadvantages of the center-tapped full wave rectifier circuit. There is no need of any center-tapping of the transformer in this circuit.

Construction

Four diodes namely D1, D2, D3 and D4 are used in constructing a bridge type network so that two of the diodes conduct for one half cycles and the other two conducts for the other half cycle of the input supply. Following figure shows the circuit diagram of a bridge FW rectifier;

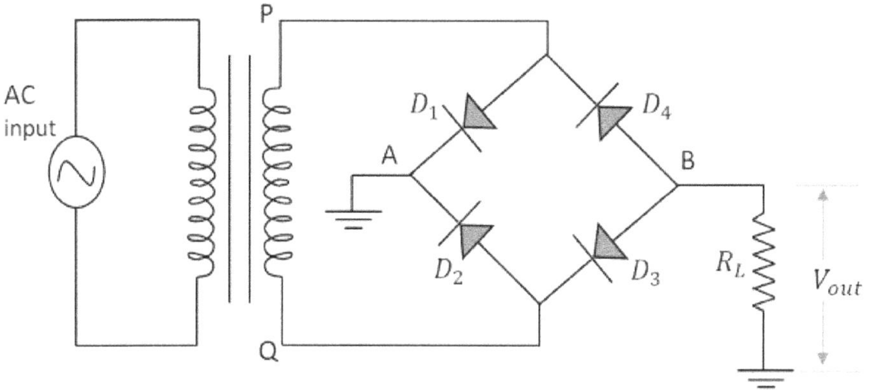

Fig: 1.12 Circuit diagram of a bridge full wave rectifier

Working

The full wave rectifier with four diodes connected in bridge circuit is employed to get a better full wave output response.

When the positive half cycle of the input supply is given, point P becomes positive with respect to the point Q. This makes the diode D1 and D3 forward biased while D2 and D4 reverse biased. These two diodes will now be in series with the load resistor R_L.

The following figure explains the working for the first half cycle;

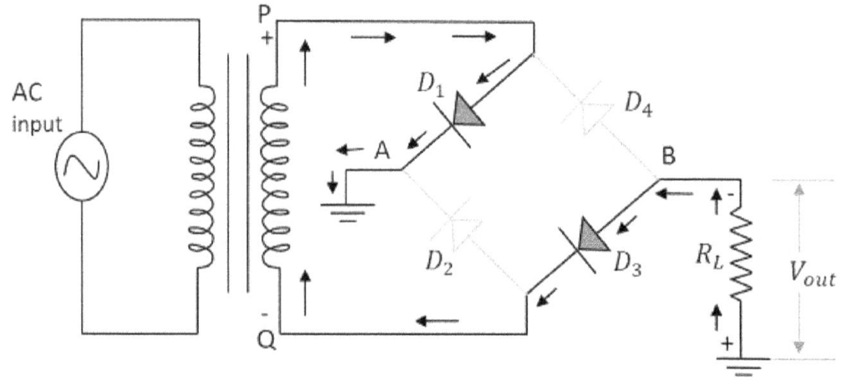

Hence, the diodes D1 and D3 conduct during the positive half cycle of the input supply to produce the output along the load resistor. As two diodes work in order to produce the output, the voltage will be twice the output voltage of the center tapped full wave rectifier.

Similarly, when the negative half cycle of the input supply is given, point P becomes negative with respect to the point Q. This makes the diode D1 and D3reverse biased while D2 and D4 forward biased. These two diodes will now be in series with the load resistor R_L.

The following figure explains the working for the second half cycle;

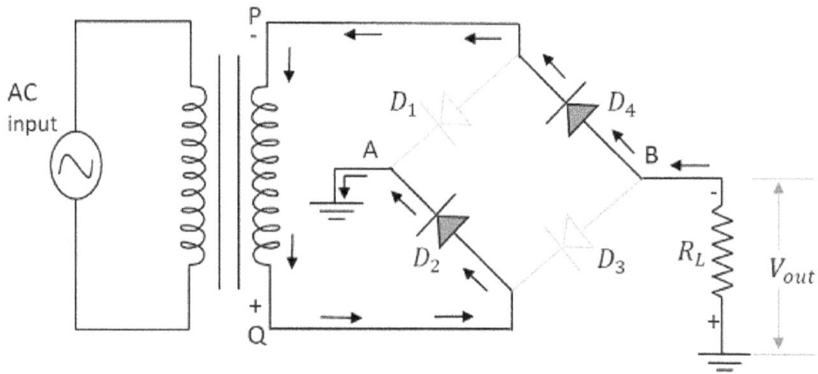

Hence the diodes D2 and D4 conduct during the negative half cycle of the input supply to produce the output along the load resistor. Here also two diodes work to produce the output voltage. The current flows in the same direction as during the positive half cycle of the input.

Waveforms

The input and output waveforms of the center-tapped full wave rectifier are given below;

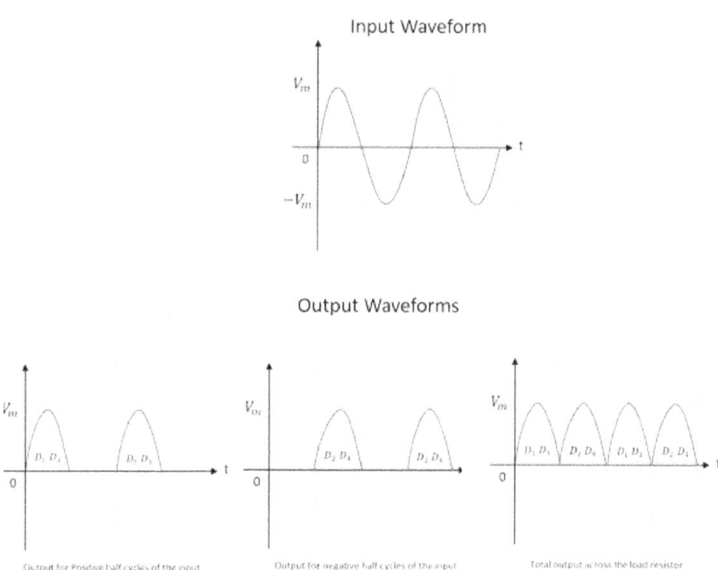

From the above figure, it is clear that the output is obtained for both the positive and negative half cycles. It can also be seen that the output across the load resistor is in the same direction for both the half cycles.

PIV

Whenever two of the diodes are placed in parallel to the secondary of the transformer, the maximum secondary voltage across the transformer appears at the non-conducting diodes, which makes peak inverse voltage of the rectifier circuit. Hence, the peak inverse voltage is the maximum voltage across the secondary winding, i.e.

$$PIV = V_m$$

Advantages of bridge FW rectifier over center tapped FW rectifier

Following are some of the advantages of a bridge FW rectifier

- No need of center-tapping.

- The dc output voltage is twice that of the center-tapper FWR.

- PIV of the diodes is of the half value that of the center-tapper FWR.

- The design of the circuit is easier with better output.

Analysis of a FW Rectifier

The total input voltage of them circuit is given by;

$V_i = V_m \sin\omega t$

Now, the current through the load resistor is given by

$$i_1 = I_m \sin\omega t \quad for \quad 0 \le \omega t \le \pi$$

$$i_1 = 0 \quad\quad for \quad \pi \le \omega t \le 2\pi$$

Where

$$I_m = \frac{V_m}{R_f + R_L}$$

R_f being the diode resistance in ON condition.

Similarly, the current i_2 flowing through diode D_2 and load resistor RL is given by,

$$i_2 = 0 \quad\quad for \quad 0 \le \omega t \le \pi$$

$$i_2 = I_m \sin\omega t \quad for \quad \pi \le \omega t \le 2\pi$$

The total current flowing through R_L is the sum of the two currents i_1 and i_2 i.e.

$$i = i_1 + i_2$$

Average output current

The average value of output current that DC ammeter read is given by;

$$I_{dc} = \frac{1}{2\pi} \int_0^{2\pi} i_1 \, d(\omega t) + \frac{1}{2\pi} \int_0^{2\pi} i_2 \, d(\omega t)$$

$$= \frac{1}{2\pi \int_0^{\pi}} I_m \sin \omega t \, d(\omega t) + 0 + 0 +$$

$$\frac{1}{2\pi} \int_0^{2\pi} I_m \sin \omega t \, d(\omega t)$$

$$= \frac{I_m}{\pi} + \frac{I_m}{\pi} = \frac{2I_m}{\pi} = 0.636 I_m$$

This value is twice from that of the half wave rectifier.

Output Voltage

The total DC output voltage across load is given by;

$$V_{dc} = I_{dc} \times R_L = \frac{2I_m R_L}{\pi} = 0.636 I_m R_L$$

This value is almost twice from the case of a HW rectifier.

I_{RMS}

The RMS value of the current is given by

$$I_{rms} = \left[\frac{1}{\pi} \int_0^\pi t^2 \, d\left(\omega t\right) \right]^{\frac{1}{2}}$$

Since current is of the two same form in the two halves

$$= \left[\frac{I_m^2}{\pi} \int_0^\pi \sin^2 \omega t \, d\left(\omega t\right) \right]^{\frac{1}{2}}$$

$$= \frac{I_m}{\sqrt{2}}$$

Rectifier Efficiency

The rectifier efficiency (η) for a full wave rectifier is given by the relation

$$\eta = \frac{P_{dc}}{P_{ac}}$$

Now,

$$P_{dc} = \left(V_{dc}\right)^2 / R_L = \left(2V_m/\pi\right)^2$$

And,

$$P_{ac} = \left(V_{rms}\right)^2 / R_L = \left(V_m/\sqrt{2}\right)^2$$

Therefore,

$$\eta = \frac{P_{dc}}{P_{ac}} = \frac{\left(2V_m/\pi\right)^2}{\left(V_m/\sqrt{2}\right)^2} = \frac{8}{\pi^2}$$

$$= 0.812 = 81.2\%$$

The rectifier efficiency can be calculated as follows –

The dc output power,

$$P_{dc} = I_{dc}^2 R_L = \frac{4I_m^2}{\pi^2} \times R_L$$

Therefore,

$$\eta = \frac{P_{dc}}{P_{ac}} = \frac{(2V_m/\pi)^2}{(V_m/\sqrt{2})^2} = \frac{8}{\pi^2}$$

$$= 0.812 = 81.2\%$$

There is an another way of calculating the rectifier efficiency;

$$P_{dc} = I_{dc}^2 R_L = \frac{4I_m^2}{\pi^2} \times R_L$$

The ac input power,

$$P_{ac} = I_{rms}^2 (R_f + R_L) = \frac{I_m^2}{2}(R_f + R_L)$$

Therefore,

$$\eta = \frac{4I_m^2 R_L/\pi^2}{I_m^2(R_f + R_L)/2} = \frac{8}{\pi^2}\frac{R_L}{(R_f + R_L)}$$

$$= \frac{0.812}{\{1 + (R_f/R_L)\}}$$

Therefore, Percentage Efficiency is

$$= \frac{0.812}{1 + (R_f + R_L)}$$

$$= 81.2\% \quad if\, R_f = 0$$

Thus, a full-wave rectifier has efficiency twice that of half wave rectifier.

Ripple Factor

The form factor of rectified output voltage of a full wave rectifier is given by;

$$F = \frac{I_{rms}}{I_{dc}} = \frac{I_m/\sqrt{2}}{2I_m/\pi} = 1.11$$

The ripple factor γ is defined as (using ac circuit theory)

$$\gamma = \left[\left(\frac{I_{rms}}{I_{dc}} \right) - 1 \right]^{\frac{1}{2}} = (F^2 - 1)^{\frac{1}{2}}$$

$$= \left[(1.11)^2 - 1 \right]^{\frac{1}{2}} = 0.48$$

Voltage Regulation

The output voltage is given by

$$V_{dc} = \frac{2I_m R_L}{\pi} = \frac{2V_m R_L}{\pi (R_f + R_L)}$$

$$= \frac{2V_m}{\pi} \left[1 - \frac{R_f}{R_f + R_L} \right] = \frac{2V_m}{\pi} - I_{dc} R_f$$

Comparison between a HW and FW rectifier

There are many features of HW and FW rectifiers that distinguish them from each other. Here, is a comparison table between both of them:

Terms	Half Wave Rectifier	Center Tapped FWR	Bridge FWR
Number of Diodes	1	2	4
Transformer tapping	No	Yes	No
Peak Inverse Voltage	V_m	$2 V_m$	$2 V_m$
Maximum Efficiency	40.6%	81.2%	81.2%
Average/dc current	I_m/π	$2I_m/\pi$	$2I_m/\pi$

Terms	Half Wave Rectifier	Center Tapped FWR	Bridge FWR
DC voltage	V_m/π	$2V_m/\pi$	$2V_m/\pi$
Ripple Factor	1.21	0.48	0.48
Output frequency	F_{in}	$2 F_{in}$	$2 F_{in}$

Applications

There are many modern day applications of rectifiers. They are widely used in several devices such as;

- DC power supplies

- Radio signals or detectors

- A source of power instead of generating current

- High-voltage direct current power transmission systems

- Several household appliances use power rectifiers to create power, like notebooks or laptops, video game systems and televisions.

1.8 Filters

The output signal that we receive after rectification still has some AC component present in it. So, in order to get a pure DC output, it is important to filter out AC component from the total output. This process is done with the help of filter circuits.

A filter circuit is a special kind of circuit which helps to remove the AC component present in the rectified output and allows only DC component to reach the load (output).

Function of a typical filter circuit could be understood by the following figure;

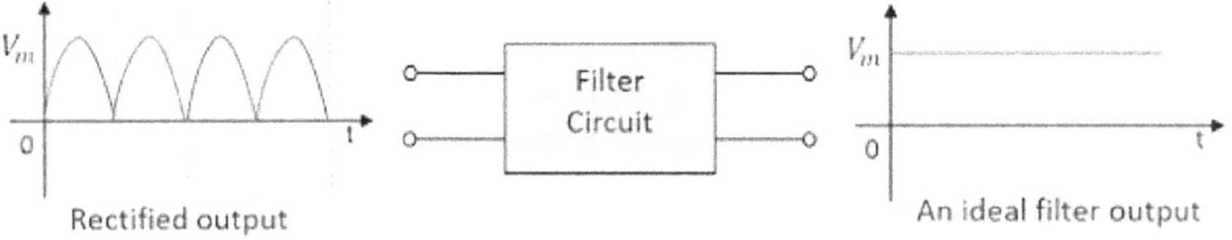

A typical filter circuit usually consists of two main components namely inductor and capacitor.

There are many types of filter circuits. Some of them are;

• Series Inductor Filter

A Series inductor filter, as the name suggests, consists of inductor connected in series, between the rectifier circuit and the load resistor. The circuit of a series inductor filter is shown in the following figure;

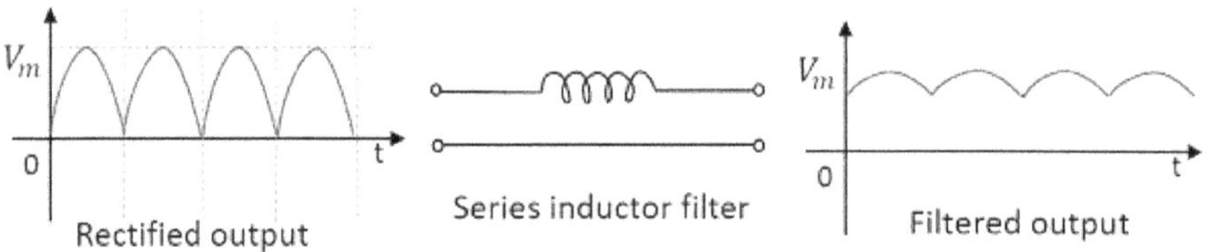

Fig:1.13 Circuit of a series inductor filter

When the rectified output is passed through the filter, the inductor blocks the AC component present in the signal, and provides a pure dc output.

• Shunt Capacitor Filter

A shunt capacitor filter consists of a capacitor connected in shunt to the rectifier and the load resistor. Following figure shows a circuit of shunt capacitor filter;

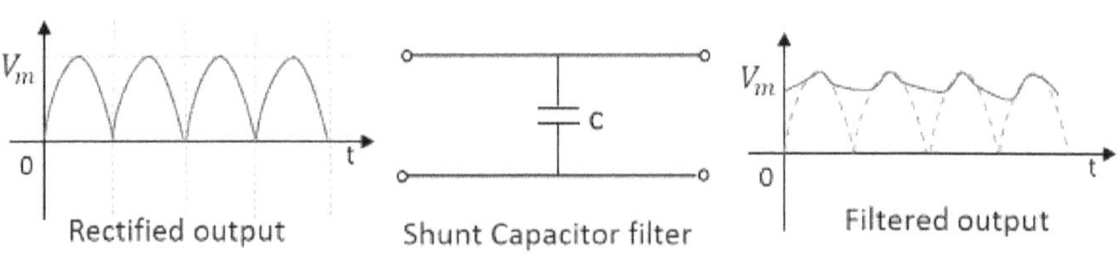

When the rectified output is passed through the filter, the AC component present in the signal get grounded through the capacitor, as the capacitor only allow the AC to pass through it. The remaining DC component present in the signal is collected at the output.

Both the series inductor filter and the shunt capacitor filter are primary type of filters. In order to improve the output quality, we can employ both of them together in different configurations. This gives rise to combinational filters.

Some important combinational filters are;

• L-C Filter

An L-C filteruses both inductor and capacitor in order to get a better output where combined efficiencies of both of them can be used. The circuit diagram of a LC filter is shown in the following figure;

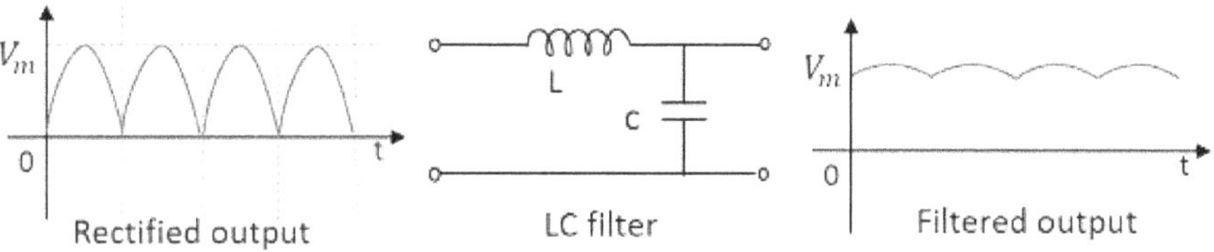

When the rectified output is given to the L-C filter circuit, the inductor allows DC component to pass through it, hindering the AC component present in the signal. Now, from that signal, more AC component present, if any, gets grounded because of the shunted capacitor, thereby giving a pure dc output.

This filter is also known as chokeinputfilter as the given input signal first enters the inductor. The output from the L-C filter is better than that of series inductor filter and shunt capacitor filter.

• ∏- Filter

∏- Filter is a commonly used type of filter circuit. It is also known as capacitor input filter since the rectified input first enters the capacitor. It consists of two capacitors and one inductor are connected in a circuit in the form of π-shaped network i.e., first a capacitor in parallel, then an inductor in series, followed by another capacitor in parallel makes this circuit. Successive identical sections can also be added according to the need of the arrangement.

The circuit diagram of π filter is shown in following figure;

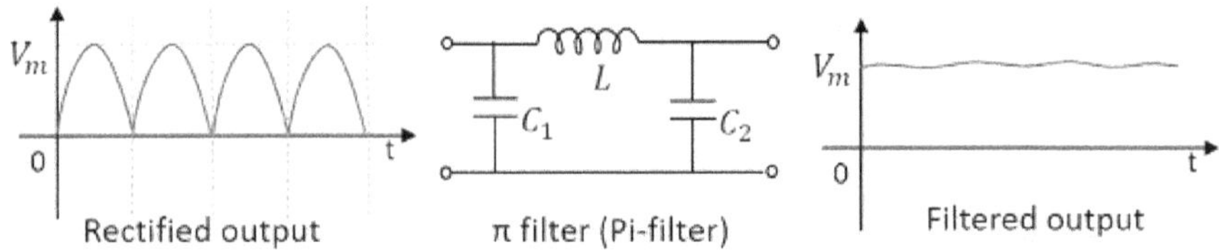

Working

In this circuit, we have two capacitors C_1 and C_2 and an inductor L. Each of them has their specific functions;

- **Capacitor C_1** – this filter capacitor offers high reactance to DC and low reactance to AC signal. After grounding the AC component present in the signal, the signal passes to the inductor for further filtration.

- **Inductor L** – This inductor offers low reactance to DC components, while blocking the AC components if any got managed to pass, through the capacitor C_1.

- **Capacitor C_2** – Now the signal is further smoothened using this capacitor so that it allows any AC component present in the signal, which the inductor has failed to block.

Thus, a pure DC output is obtained.

1.9 Zener Diode as Voltage Regulator

A zener diode is one of the specially designed diodes that predominately work in reverse biased conditions. They are more heavily doped than ordinary diodes, due to which they have narrow depletion region. While regular diodes get damaged when the voltage across them exceeds the reverse breakdown voltage, Zener diodes work exclusively in this region. The depletion region in Zener diode goes back to its normal state when the reverse voltage gets removed. This particular property of Zener diodes makes it useful as a **voltage regulator**. Let us see how this happens.

Working of a Zener Diode

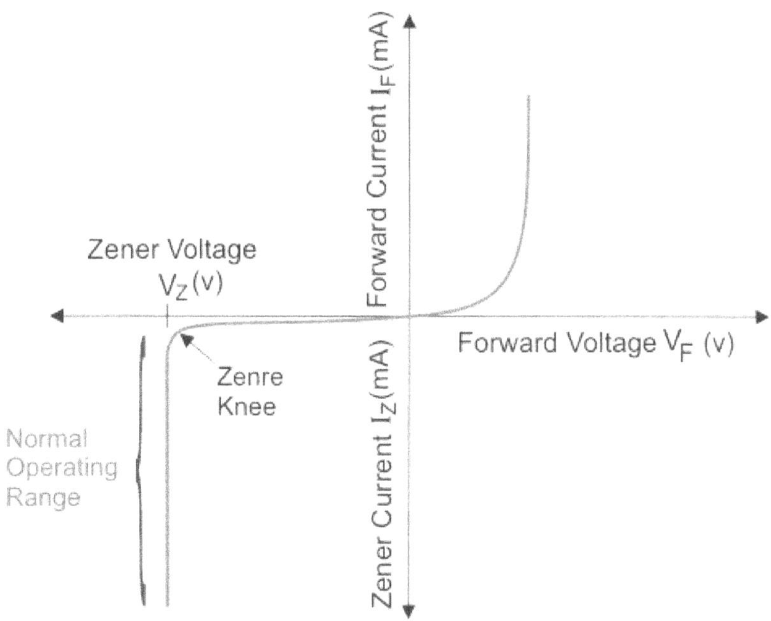

Fig:1.14 V-I characteristics for a Zener diode

When we apply a reverse voltage to a Zener diode, a negligible amount of current flows through the circuit. When a voltage higher than Zener breakdown voltage is applied, zener breakdown occurs. Zener breakdown is a phenomenon where a significant amount of current flows through the diode with a negligible drop in voltage. When we increase the reverse voltage further, the voltage across the diode remains at the same value of Zener breakdown voltage whereas the current through it keeps on rising as seen in the graph above. Here in the graph V_z refers to the Zener breakdown voltage. Zener breakdown voltage typically can range from 1.2 V to 200 V depending on its application.

The exciting part of this diode is that we can choose the zener diode with a suitable breakdown voltage to work as a voltage regulator in our circuit.

For example, we want that the voltage across a load in our circuit does not exceed, let's say, 12 volts. Then we can select a Zener diode with a breakdown voltage of 12 volts and connect it across the load. Then even if the input voltage exceeds that value, the voltage across the load will never exceed 12 volts.

Let's try to understand that with a circuit diagram;

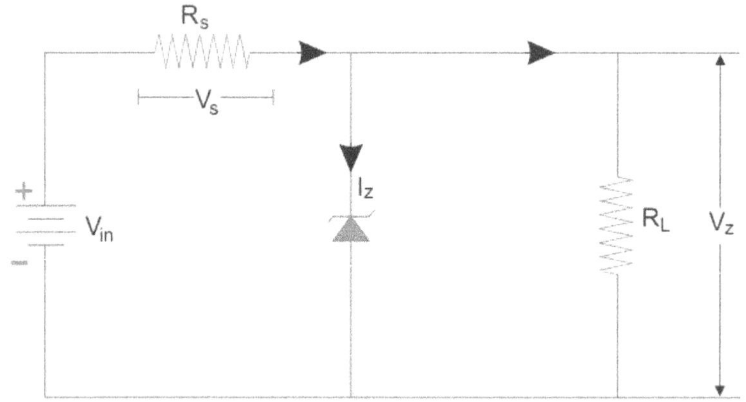

Fig:1.15 Circuit diagram for a zener diode

Here, the Zener diode is connected across the load R_L. We want the voltage across the load to be regulated and not cross the value of V_Z. Depending on our requirement; we choose the suitable Zener diode with a Zener breakdown voltage near to the voltage we require across the load. We connect the Zener diode in reverse bias condition. When the voltage across the diode exceeds the Zener breakdown voltage, a significant amount of current starts flowing through the diode. As the load is in parallel to the diode, the voltage drop across the load is also equal to the Zener breakdown voltage. The Zener diode provides a path for the current to flow and hence the load gets protected from excessive currents. Thus, the Zener diode serves two purposes here: **Zener diode as a voltage regulator** as well as it protects the load from excessive current.

SUMMARY

- A diode is equipment which allows the electric current to pass in one direction only. In the other direction it produces a very high resistance thereby hindering the current to flow in reverse direction.

- Diode can convert AC signal to DC signal therefore can used diode rectifiers.

- When a diode or any two-terminal component is connected in a circuit, it has two biased conditions with the given supply. They are Forward biased condition and Reverse biased condition.

- A ideal diode is a two-terminal electronic component that conducts current primarily in only one direction i.e., it has very low (ideally zero) resistance in one direction, and very large (ideally infinite) resistance in the other.

- A clipper is a device that removes or limits some portion of the input wave form (input voltage signal) above or below a certain given level.

- A Clamper circuit is a combinational circuit of a diode, a resistor and a capacitor that transforms the input waveform to a desired DC level without altering the applied signal. It actually additions a DC level to an AC signal.

- A rectifier is simply electrical equipment that converts alternating current (AC) to direct current (DC). It is generally composed of one or more diodes.

- A filter circuit is a special kind of circuit which helps to remove the AC component present in the rectified output and allows only DC component to reach the load (output).

- A zener diode is one of the specially designed diodes that predominately work in reverse biased conditions.

SHORT ANSWER TYPE QUESTION

Q. 1.   Draw a characteristic curve for a p.n junction diode?

Q. 2.   What happens to the width of depletion layer when a p-n junction diode is forward- biased ? What when reverse-biased?

Q. 3.   Why does the width of the depletion increase when the p-n junction is reverse-biased?

Q. 4.   The small reverse current Mowing through a reverse-biased junction diode is called the 'saturation' current. Why?

Q. 5.   A Si diode has a much smaller reverse current than a Ge diode of the same size. Why?

Q. 6.   What is surface-leakage current in a junction diode?

Chapter 2

Bipolar Junction Transistors

A Brief Chapter Overview

BIPOLAR JUNCTION TRANSISTORS

2.1 Introduction

Being regarded as one of greatest invention of 20th century, transistors became the founding chapter for modern electronics.

During early 1900, vacuum tubes were the only technology in all electronics. Now outdated, these vacuum tubes uses phenomenon of thermo-ionic emission of electrons from a heated cathode. A cathode heater consumed significant amount of power, so heavy power supply was required for the equipment. Functional delay was also noticed as tube heaters require time to warm up. Also due to cathode heating a whole lot of heat was produced which circuit designers had to take care about. Thus the limited technology results into heavy and outdated equipment that once dominated the whole industry.

But it was in 1948, an American inventor named William Shockley from Bell Labs came up with an invention that revolutionized the whole industry. The invention was Bipolar Junction transistor (BJT) which later on became an active component of all modern electronics.

Transistors immediately replaced ancient vacuum tubes as they were power efficient and flexible in nature.

Commercialization of new transistor designs led to beginning of a trillion-dollar semiconductor industry and the California's Silicon Valley which became the universal hub for technology.

Later on dominating for three decades, simple transistors were replaced by IC's and CMOS technology was a direct derivative of modified transistor technology.

2.2 Some Basic Ideas

A simple Bipolar Junction Transistor is basically made of two PN-junction diodes thereby forming three connecting terminals. Each terminal is specifically named according to their respective properties. These three terminals are known as the Emitter (E), the Base (B) and the Collector (C) respectively.

As the name suggests, The EMITTER emits or injects the electrons to the base terminal. This is the reason why the emitter is so heavily doped, so that it can have sufficient charges to be injected. The BASE region acts as a bridge between the emitter and the collector regions. It is lightly doped and has very less thickness thereby letting most of the negative charges from the emitter to the collector terminal. The third one is the COLLECTOR. Collector is so named because it accumulates or collects electrons from the base terminal. It is the largest of the three regions and it dissipates more heat than the emitter or the base terminal. Collector is doped somewhere between the highly doped emitter and the lightly doped base.

BJTs are current regulating devices, that is, they control the amount of current flowing through them in accordance with the amount of biased voltage applied to their base terminal. The principle of operation of the two types of transistors namely N-P-N and P-N-P, is exactly the same the only difference being in their biasing and the polarity of the power supply for each type.

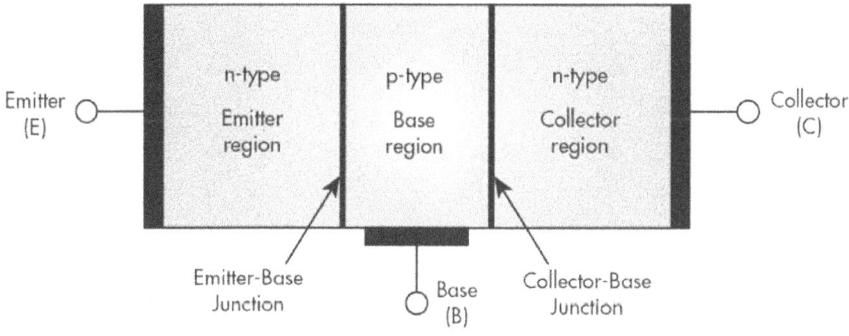

Figure: 2.1 An unbiased NPN transistor with different components

UNBIASED TRANSISTOR

What is an unbiased transistor?

The state at which no external voltage, i.e., not any sort of biasing is applied on the transistor, is called unbiased state of that transistor.

So, there will be no current flowing through any of the transistor terminals. This is the equilibrium state of the transistor.

Since, a BJT is a combination of two p-n junction diodes compiled with each other, there is the formation of depletion regions at both the transistor junctions, i.e., EMITTER-BASE junction and COLLECTOR-BASE junction, as shown in the figure. These depletion regions, during diffusion, penetrates more deeply into the lightly doped base so that the number of impurity atoms remains the same on either side of the junction.

The penetration by the depletion region at emitter side of the junction is much less than the base side of the junction because of the heavy doping of the emitter region. Similar is the case at the collector junction. Since the collector is intermediately doped, the depletion width at the collector junction is more as compared to the depletion width at the emitter junction.

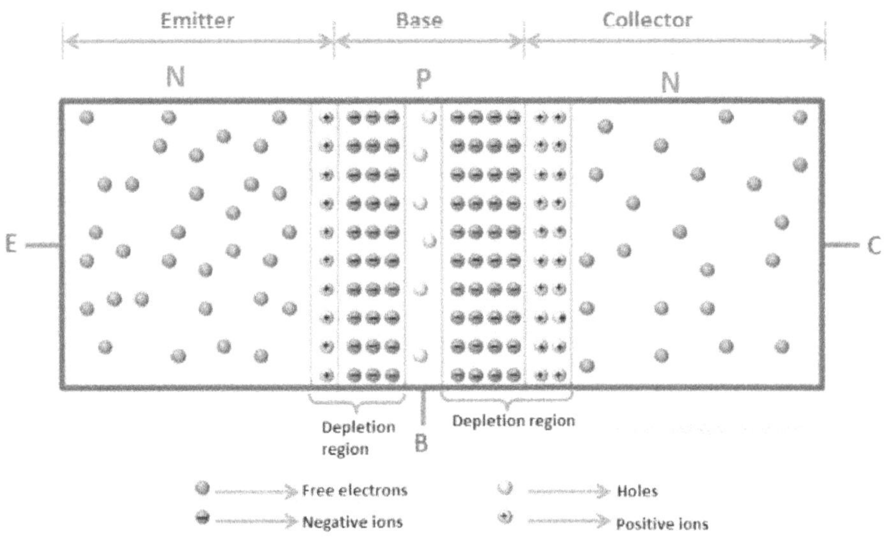

Figure : 2.2 Constituents of a typical transistor

TRANSISTOR BIASING

What biasing means?

Bias is "a potential applied to a device (such as a transistor, diode etc.) to establish a reference point or level for operation". A transistor bias, however, can also be a current; not only a voltage.

A bipolar junction transistor (BJT) without external bias is non-conducting, and will typically begin conducting between its collector and emitter region when a relatively small amount of current is introduced into the base-emitter junction.

There are several ways to connect and to operate a transistor, depending on the intended application, and many modes of biasing as well.

A BJT transistor can be run with:

Zero bias: In RF power applications, this is called Class C biasing; the transistor conducts for less than half of the signal waveform.

Slight forward bias: In linear amplifier applications, this is called Class B biasing; the transistor is conducting for roughly 1/2 of the input signal waveform.

Forward bias: In linear amplifier applications, this is called Class A biasing; the transistor is conducting 100% of the input signal waveform.

A transistor can be made to operate such that it is conducting only for a small fraction of time: for instance, only on the peaks of the signal.

The base-emitter junction is mostly negatively biased.

BIASING: FORWARD AND REVERSE

Forward Biased P-N Junction

When p-type region of a junction is connected to the positive end and n-type region is connected to the negative end of a battery (voltage source), then the junction is said to be forward biased.

During forward biasing, ē involved in the covalent formation in the p-type semiconductor is attracted towards the positive end of the battery, thereby resulting in the breakage of these covalent bonds. These covalent bonds supply è during biasing. As the number of broken bonds increases, more and more ē is slipped towards the positive end of the source which results in increase in concentration of the electrons near the terminal. Here, this ē recombined with holes which are already there. Thus, the concentration of holes increases in the p-type region's portion away from the junction and decreases in the portion nearer to the junction. Due to terminal shift of ē, holes are also shifted from the terminal towards the junction.

Higher concentration of holes adjacent to the impurity ions makes the electrons from these impurity ions recombine with the holes forming fresh holes in the layer. This results in the reduction of width of negative ions layer and finally the layer vanishes.

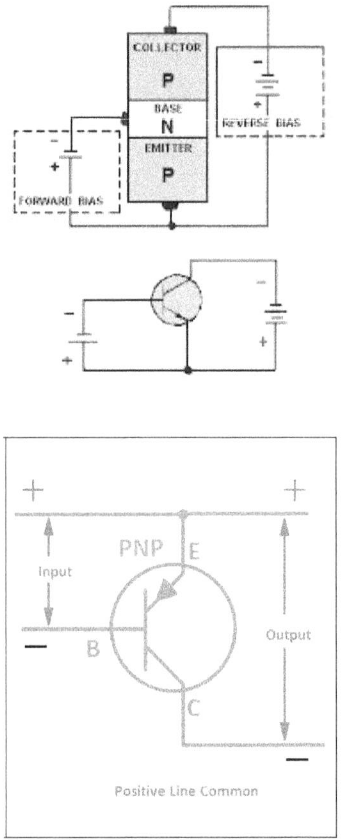

Figure: 2.3 A typical PNP transistor in biasing

Similar is the case at the other side. The free ē in the n-type region begins to repel from the terminal towards the junction due to the negative end of the battery. This leads to a higher concentration of free ē near the junction. Here, they recombine with the layer of positive impurity ions, already present there, and gives rise to more fresh free ē inside the layer. As a consequence, the width of the layer of positive impurity ions starts getting reduced and finally vanishes.

Now, since both the layers of impurity ions are vanished on either side of the junction, there will be no more depletion layer left.

After the depletion layer is completely disappeared, holes can easily move from the p-type to the n-type region and free ē from the n-type to the p-type region in the semiconductor. Therefore, there will be no hindrance to the current flow and thus, the p-n junction acts as a short circuit.

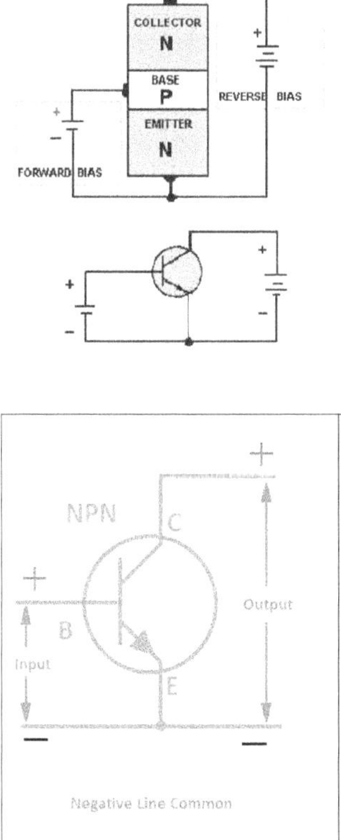

Figure: 2.4 A typical NPN transistor in biasing

Forward Biased P-N Junction

When p-type region of a junction is connected to the positive end and n-type region is connected to the negative end of a battery (voltage source), then the junction is said to be forward biased.

Reverse Biased P-N Junction

A P-N junction is said to be reverse biased, when the n-type region is connected to the positive end and p-type region is connected to the negative end of the battery (voltage source).

When there is no EMF applied across the junction, there exist some natural potential (V) across the junction forming a potential barrier. Values of V for some different semiconductors at room temperature (25°C) at no external voltage;

PN JUNCTION	Potential (V)
Silicon(Si)	0.7
Germanium(Ge)	0.3

Due to the higher value for the potential barrier, silicon is preferred over germanium in modern day semiconductor devices.

The potential barrier has the same polarity as the voltage source in reverse biased condition. Therefore, if the reverse voltage across the p-n junction is increased, then the potential barrier across that junction will also tend to increase. When the n-type region is connected to the positive end of the battery, the free ē from that region is shifted towards the positive end due to attraction. This leads to the generation of more positive ions in the depletion region that makes the positive ions layer thicker. At same instant, due to the connection of p-type region with the negative end of the battery, fresh ē is injected into this region.

Since, the n-type region has gained some positive potential at the junction, e⁻ moves towards the junction and recombine with holes present adjacent to the positive ion layer and create more and more positive ions in the layer. This overall leads to the thickening of the layer.

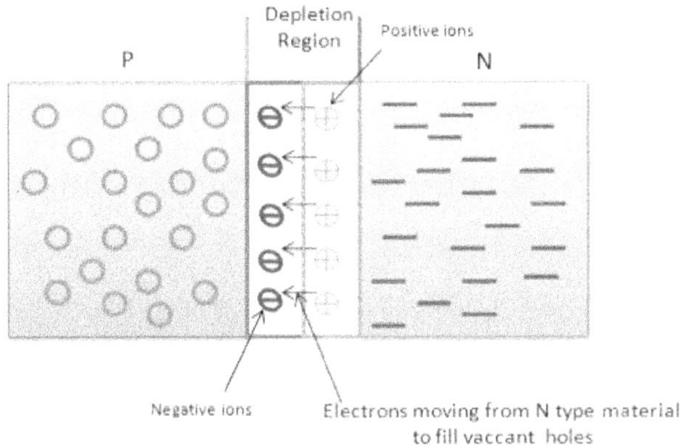

Figure: 2.5 Depletion region between P-N regions

Hence, the total depletion width increases along with its barrier potential and tends to increase until this barrier potential reaches the value of the applied reverse voltage. Normally, this increase in the barrier potential continues until it reaches the applied reverse voltage but if the reverse voltage is sufficiently high then the depletion layer will vanish. This happens normally under two different types of phenomenon:

- Zener breakdown

- Avalanche breakdown

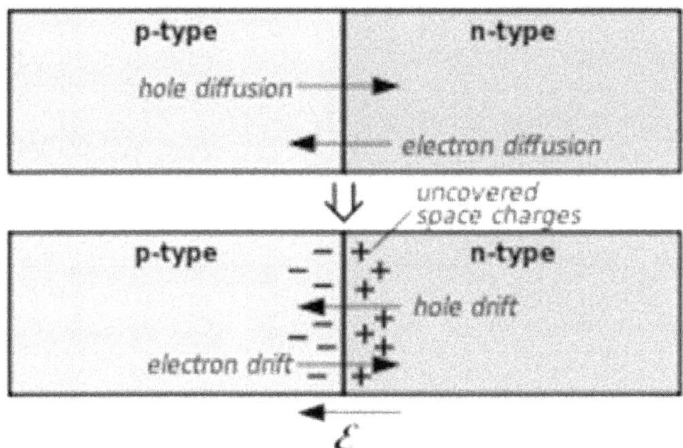

Figure : 2.6 Charge carrier flow in a typical transistor

If the reverse voltage is not sufficiently high, then the barrier potential reach the value of the reverse voltage and the completion of depletion layer takes place. At this point, there is no drift of the charge carriers across the junction because the potential barrier nullifies the effect of applied voltage as both of them have the same value.

But there is some tiny current from n-type to p-type region because of thermally generated ē and holes.

Forward Current in PN Junction

When the PN junction is forward biased i.e., P-type region is connected to the positive terminal and N-type region is connected to the negative terminal of the battery which causes a continuous flow of charges through the junction constituting a current known as FORWARD BIASED CURRENT. The forward current i_D is given by the equation;

$$Forward\ biased\ current\ \ i_D = I_s \left(e^{\frac{v_D}{nV_T}} - 1 \right)$$

Where,

I_S = Saturation Current (10^{-9} to 10^{-18} A)

V_T = Volt-equivalent temperature (= 26 mV at room temperature)

n = Emission coefficient (1 ≤ n ≤ 2 for Si ICs)

This equation is an approximation for forward current.

Reverse Current in PN Junction

When the junction is reverse biased i.e., its n-type region is connected to the positive potency of the battery and the p-type region is connected to the negative potency of the battery, we ideally assume that no current is flowing across the junction. But this is not technically correct.

A tiny amount of current is there due to the thermally generated ē and holes. This is called the reverse current generally denoted by i_0. i_0 has very small value($i_0 \approx 0$).

i_0 can be expressed by the relation:

$$Reverse\ biased\ current\ \ i_0 \approx I_s e^{\frac{V_D}{nV_T}}$$

I_S = Saturation Current (10^{-9} to 10^{-18} A)

V_T = Volt-equivalent temperature (= 26 mV at room temperature)

n = Emission coefficient ($1 \leq n \leq 2$ for Si ICs)

Specifications of a typical PN Junction

A p-n junction is specified in four manners.

1. Forward Voltage Drop (V_F): Is the forward biasing junction level voltage (0.3V for Germanium and 0.7V for Silicon Diode)

2. Average Forward Current (I_F): It is the forward biased current due to the drift electron flow or the majority carriers. If the average forward current exceeds its value the diode gets over heated and may be damaged.

3. Peak Reverse Voltage (V_R): It is the maximum reverse voltage across the diode at its reverse biased condition. Over this reverse voltage diode will go for breakdown due to its minority carriers.

4. Maximum Power Dissipation (P): It is the product of the forward current and the forward voltage.

V-I Characteristics of a PN Junction

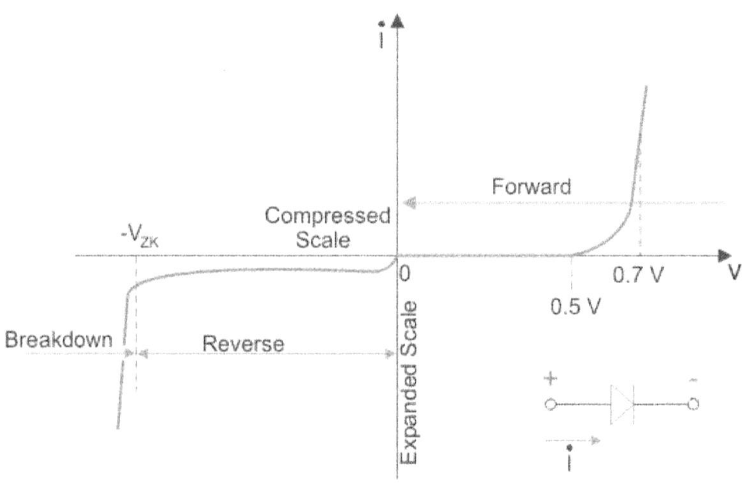

Figure : 2.7 V-I characteristics curve for a PN junction

In the forward bias, the operational region is in the first quadrant. The threshold voltage for Germanium is 0.3 V and for Silicon is 0.7 V. Beyond this threshold voltage the graph goes upward in a non linear manner. This graph is for the dynamic Resistance of the junction in the forward bias.

In the reverse bias voltage increases in the reverse direction across the p-n junction, but no current due to the majority carriers, only a very small leakage current flows. But at a certain reverse voltage p-n junction breaks in conduction. It is only due to the minority carriers. This amount of voltage is sufficient for these minority carriers to break the depletion region. At this situation sharp current will flow through this junction. This breakdown of voltage is of two types.

- **Avalanche Breakdown:** it is not properly sharp, rather inclined linear graph i.e. after break down small increase in reverse voltage causes more sharp current gradually.

- **Zener Breakdown:** This breakdown is sharp and no need to increase reverse bias voltage to get more current, because current flows sharply.

2.3 BJT on Energy Basis

A better understanding of transistors can be done through their energy diagrams. Figure shows the energy diagram of an NPN transistor. Here the emitter diode is forward biased which makes the free ē in the emitter region to move to the base region. Why this happens? It happens because now the orbits of some emitter ē are big enough to match the available base orbits. This leads to the diffusion of these emitter electrons from the emitter conduction band to the base conduction band.

Now, when these emitter electrons enter the base conduction band, they become minority charge carriers as they are inside a P-region (holes are majority charge carriers here). This makes the base a highly dense minority carrier region. In any typical transistor, nearly 95 percent of these minority carriers have a lifetime long enough to diffuse into the depletion layer of collector and fall down the collector energy hill. When they fall, they lose energy in the form of heat. The collector must be capable enough to dissipate this heat energy which is the reason why the collector is largest of the three regions. The remaining emitter region injected electrons (nearly 5 percent) fall along the recombination path as shown in the figure. Those that do recombine become valence ē and flow through base holes into the external base lead.

Conclusion:

- The forward biasing of the emitter diode controls the frequency of free ē injected into the base region. Larger the base-emitter voltage (Vbe), more the number of injected ē.

- The reverse biasing of the collector diode affects very little the frequency of ē entering the collector region. Increasing collector-base voltage may steepen the collector energy hill but this doesn't majorly change the number of free ē entering the collector depletion layer.

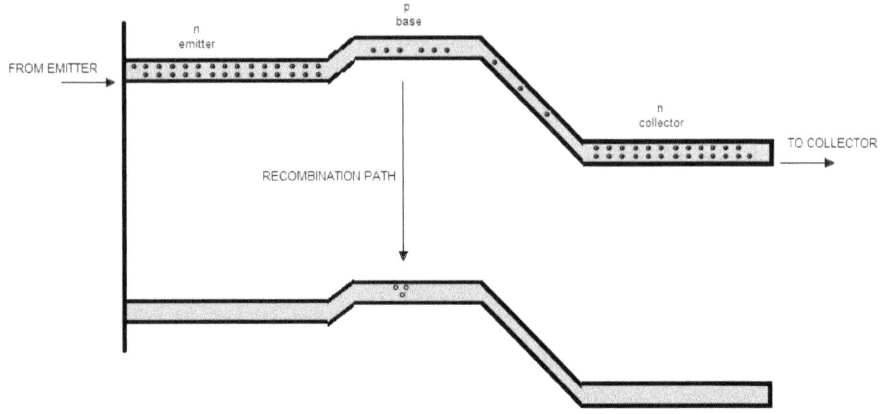

Figure : 2.8 Energy bands of a typical NPN transistor

CURRENT COMPARISON (α_{DC})

α_{DC} of a transistor draws a relation between the transistor current values. If more than 95 percent of injected ē reaches the collector, this means that the collector current(I_C) is almost equal to the emitter current (I_E). This is related as α_{DC},

$$\alpha_{DC} = I_C/I_E$$

2.4 Transistor Characteristics

There are three possible ways to connect BJT with an electronic circuit as the Bipolar Junction Transistor is a three terminal device, with one terminal being common to both the input and output. Each method of connection has different input signal with output signal and forms characteristics of transistor configuration

1. Common Base Configuration (CB)

2. Common Emitter Configuration (CE)

3. Common Collector Configuration (CC)

Although CC configuration exists, we will confine our study only to CB and CE configuration as they of main importance.

2.5 Common Base Configuration

In Common Base (CB) or grounded base configuration, the base terminal is common to both the input and the output signal. The input signal is applied between base and emitter terminals and the corresponding output voltage is taken from the collector terminal with the base terminal grounded or connected to a voltage terminal from where all the voltages are taken. Circuit diagrams of CB configuration of PNP and NPN transistors are represented by the figure (a) and (b) respectively.

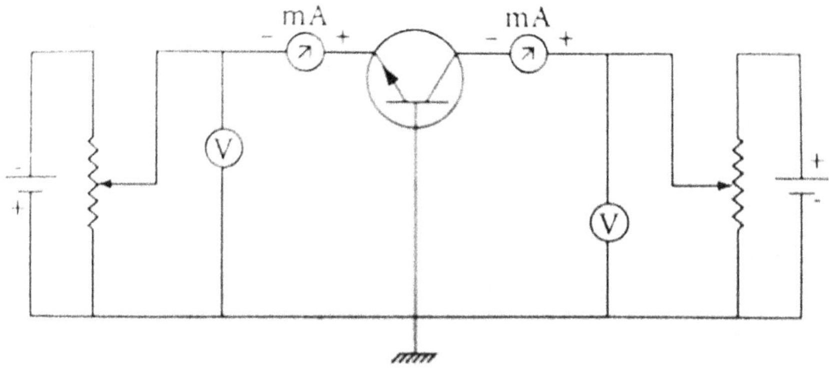

Figure (a): 2.9 Circuit diagram for CB configuration of a NPN transistor

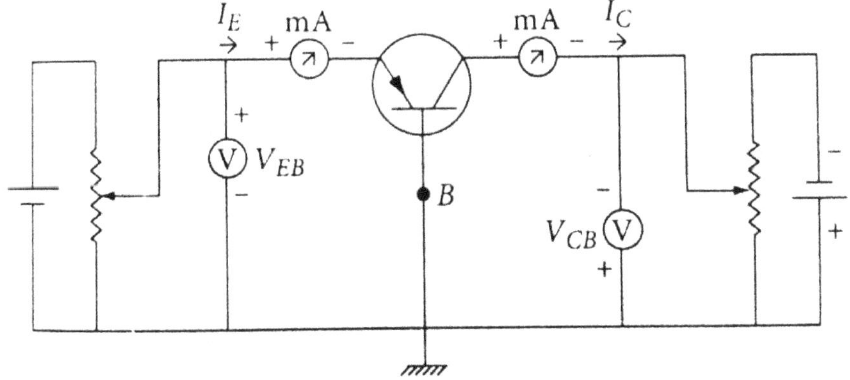

Figure (b): 2.10 Circuit diagram for CB configuration of a PNP transistor

For a PNP transistor I_E is positive, while I_B and I_C are negative. Input voltage (V_{EB}) is positive for the forward biased emitter junction while output voltage (V_{CB}) is negative for its reverse biased collector junction. There are two set of characteristic curve that completely depicts the behaviour of CB configuration; one for the input signal and other for the output signal.

The input characteristics curve for CB configuration shows the alteration of input current I_E with input voltage (V_{EB}) for various values of output voltage (V_{CB}).

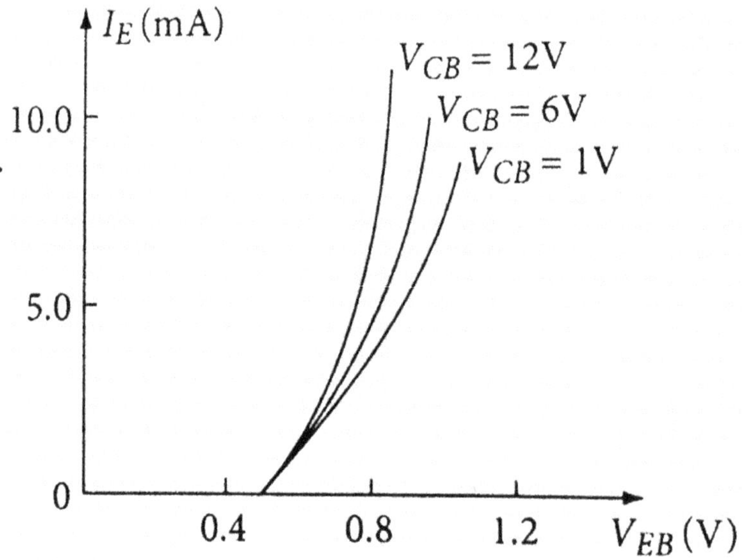

Figure : 2.11 Input characteristics curves for a CB configured transistor

Input characteristics curves are shown in figure. Following are the keys points that are depicted by this figure:

1. For the fixed values of output voltage, the emitter current I_E increases with increase in input voltage (V_{EB}), in a way similar to that of a PN junction diode;

2. For a fixed input voltage (V_{EB}), emitter current I_E increases with increase in reverse collector voltage (V_{CB}) which however is very small in nature and in most cases is generally neglected.

The output characteristic curve for CB configuration shows the alteration of output current I_C with output voltage (V_{CB}) for several of input current I_E.

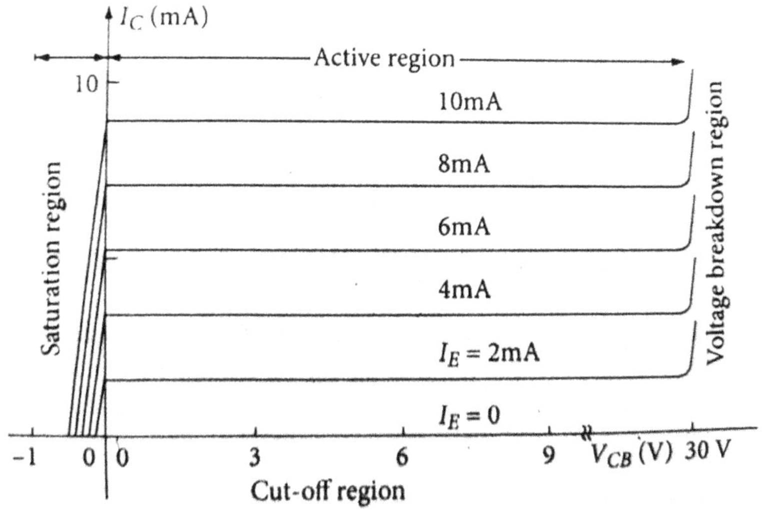

Figure: 2.12 Output characteristics curves for a CB configured transistor

The output characteristic curves are shown in figure. From the figure, it can be seen that the whole plot of curves is divided into three regions which explains the output characteristic nature of CB configuration. These regions are;

• ACTIVE REGION

In active region of the plot, the emitter-base junction is forward biased and the collector-base junction is reverse biased. For its region close to that of the abscissa, $I_E = 0$, and the output current I_C is equivalent to reverse saturation current (I_{C0}) As I_{C0} is very tiny in magnitude, curve for I_E overlaps with the x-axis (voltage axis).

In the active region, I_C is independent of output voltage V_{CB} as the output current I_C appear as a horizontal line. But, in reality there is a very minute slope in output current with V_{CB} due to the early effect. However this variation is neglected in most cases. This active region is essentially linear and employee for linear amplification.

• SATURATION REGION

In saturation region of the plot, both emitter-base junction and collector-base junction are forward biased; thereby depicting the characteristics curve in the left of current axis.

When the collector junction is forward biased, fresh holes are injected from P-type collector to N-type base, thereby reducing the collector current rapidly(I_C may go negative). In this region, output current I_C increases exponentially with collector voltage V_{CB}

• CUT OFF REGION

In the cut-off region, both emitter-base junction and collector-base junction is reverse biased. From the plot, it can be seen that I_C is zero in this region. That is why it is called the cut-off region. In this region of action, only reverse saturation current (I_{C0}) flows through the collector.

Alpha (α)

As described earlier, α is defined as ratio of change in the collector current from cut-off to emitter current change from cut-off. α is called large signal current gain of transistor in CB configuration.

$$\alpha_{dc} = I_C / I_E$$

For commercial transistors, the value for α lie in the range of 0.90 – 0.995.

The common base configuration is mostly used in circuits such as radio frequency (RF) single stage amplifier circuits or microphone pre-amplifier due to its very good high frequency response.

2.6 Common Emitter Configuration

Mainly, the transistor circuits that we normally take into use are of common emitter configuration. It is also sometimes called grounded emitter configuration because the emitter is common to input and output both.

The circuit diagram for a common emitter (CE) configuration of a PNP and a NPN transistor is shown in Figure (a) and (b) respectively.

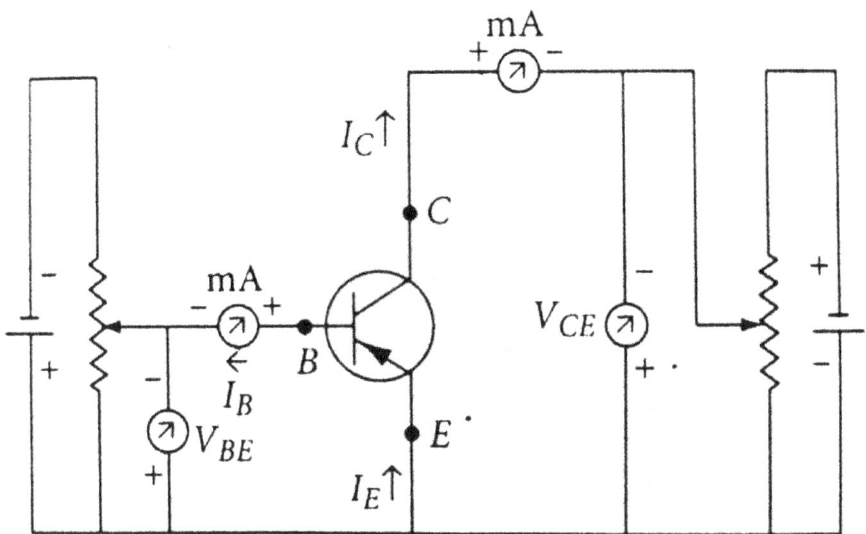

Figure (a): 2.13 Circuit diagram for CE configuration of a PNP transistor

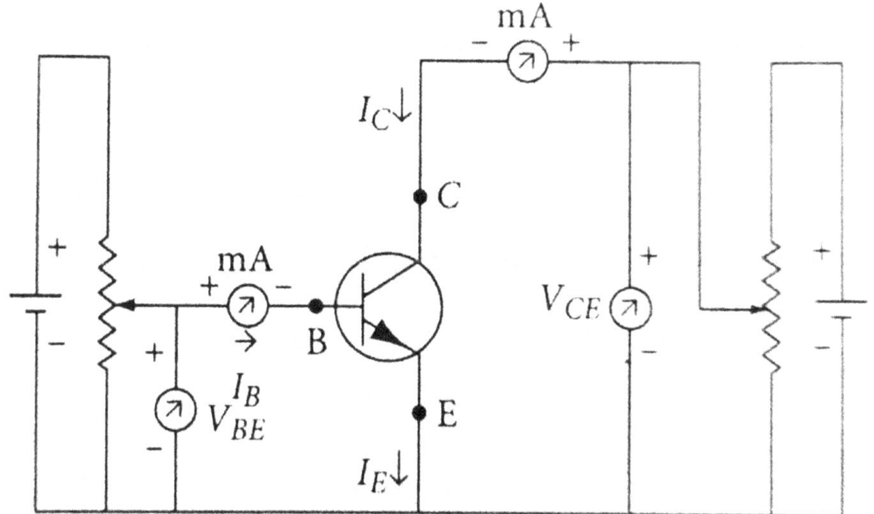

Figure (a): 2.14 Circuit diagram for CE configuration of a NPN transistor

The behaviour of a CE configured transistor can be completely understood by its input and output characteristics curves. Input characteristics curve for CE configuration describes the variation of input current (I_B) verses input voltage (V_{BE}) for various values of output voltage V_{CE}.

Input characteristics curves are shown in figure. Following are the keys points that are depicted by this figure;

1. For fixed output voltage V_{CE}, input current rises exponentially with V_{BE}, when V_{BE} exceeds the cut in voltage.

2. For fixed input voltage V_{BE}, input current I_B decrease with increase in output voltage V_{CE}. This is because with increases in collector-base junction, reverse bias voltage of effective base width decreases, so recombination in base region is less and I_B decreases. However, this alteration of I_B with V_{CE} for fixed V_{BE} is very small, so we neglect this for simple use.

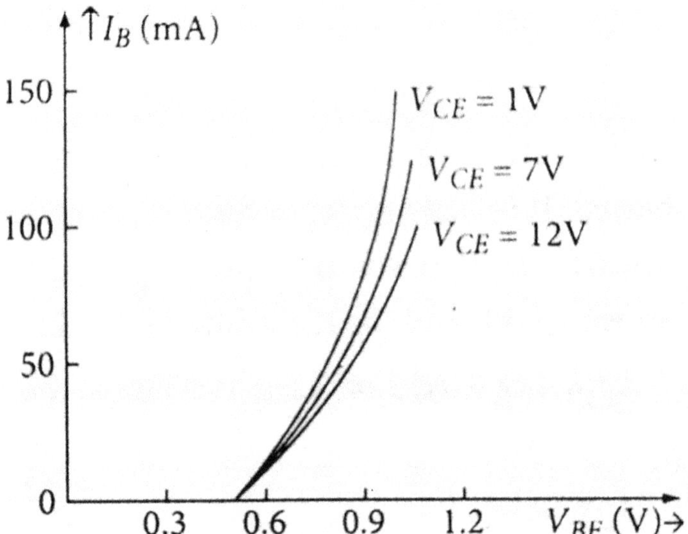

Figure: 2.15 Input characteristics of a transistor in CE configuration

The output characteristics curves for the CE configuration describes the variation of output current (I_C) verses output voltage (V_{CE}) for a range of values of input current I_B, shown in the figure.

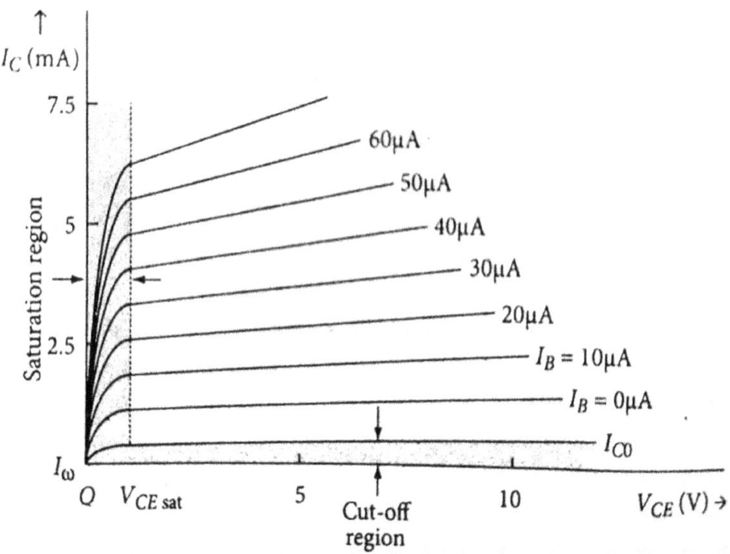

Figure: 2.16 Output characteristics of a transistor in CE configuration

Following are the key observations from the plot;

- For fixed values of I_B, the collector current I_C increases rapidly with increases of V_{CE} value from 0V to V_{CEsat} (saturation voltage), the collector current I_C does not vary widely with the collector voltage V_{CE}, when V_{CE} just above a small voltage V_{CEsat}. However, the curves for different values of I_B are not horizontal as obtained for different values of I_E in CB configuration.

- It can also be noted from this figure that I_B is much smaller than I_E.

From the figure, it can be seen that the whole plot of curves is divided into three regions which explains the output characteristic nature of CE configuration. These regions are;

1. ACTIVE REGION

2. SATURATION REGION

3. CUT-OFF REGION

From the figure it can be seen that no polarity change of the signal is necessary for switching from saturation region to the active region.

1. ACTIVE REGION

In active region, the emitter junction is forward biased and the collector junction is reverses biased. In this region of output characteristics, the characteristics curves have greatest linearity and curves are equally spaced for equal separation of I_B. In the figure, it is the region above $I_B = 0$ and the right side of the dashed line at V_{CEsat}.

BETA (β)

In CB configuration, I_C and I_B are related to each other by a quantity called β, and is given by the relation;

$$\beta_{dc} = I_C / I_B$$

β is the large signal current gain in CE configuration.

2. CUT-OFF REGION

The cut-off of a CE configured transistor is defined as $I_E = 0$, $I_C = -I_B$ and V_{BE} is reverse voltage of order 0.1V and 0V for germanium and silicon respectively.

3. SATURATION REGION

Both emitter and collector junctions are forward biased in the saturation region. In figure, the saturation region is that of the left of the dotted line through V_{CEsat} and the y-axis through $V_{CE} = 0$.

This type of transistor configuration is used most in circuit for transistor based amplifiers.

RELATION BETWEEN α_{dc} and β_{dc}

Kirchhoff's current law gives us,

$$I_E = I_C + I_B$$

This tells us that emitter current is the sum of Collector current and base current. It can always be noted that the emitter current is largest of the three currents, collector current is almost as large, and base current is much smaller.

Dividing both sides of eq. by IC,

$I_E/I_C = 1 + I_B/I_C$

or,

$1/\alpha_{dc} = 1 + 1/\beta_{dc}$

By rearranging, we get

$\beta_{dc} = \alpha_{dc}/1-\alpha_{dc}$

Occasionally, we need a formula for α_{dc} in terms of β_{dc}. So we can rearrange eq. to get,

$\alpha_{dc} = \beta_{dc}/\beta_{dc}+1$

Load line analysis

Load line is the locus of operating point of a transistor. It is described by a graph plotted between I_C and V_{CE} for any transistor configuration. It helps to determine respective values of collector current and collector to emitter voltage to operate transistor in any particular mode or region.

From earlier observations, we found that the transistor operates well in active region and hence, it is also known as linear region. The outputs of the transistor are the collector current and collector voltages.

When the output characteristics of a transistor are considered, the curve looks as below for different input values.

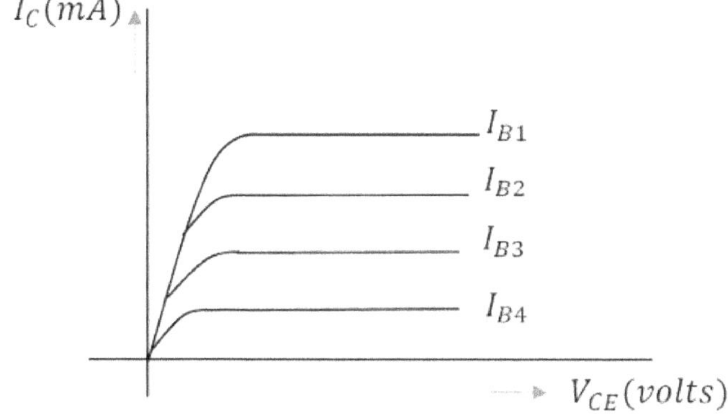

Figure : 2.17 Output characteristics curve for a typical transistor

From the above figure, we can see that output characteristics are drawn between collector current I_C and collector voltage V_{CE} for different values of I_B. These are measured here for different input values to obtain different output curves.

Operating point

The maximum possible value for the collector current can be defined by a point present on the Y-axis. This is known as saturation point. Similarly, the maximum possible value for collector-emitter voltage can be defined by a point on the x-axis. That point is called cut-off point.

When a line is drawn joining these two points, i.e., the saturation point and the cut-off point, that line is called as **Load line**. This is so named because it defines the output at the load. This line, when drawn over output characteristic curve, makes contact at a point known as **Operating point**. This operating point is also called as quiescent point or **Q-point**.

There can be many such intersecting points, but the Q-point is selected in such a way that irrespective of AC signal swing, the transistor remains in active region.

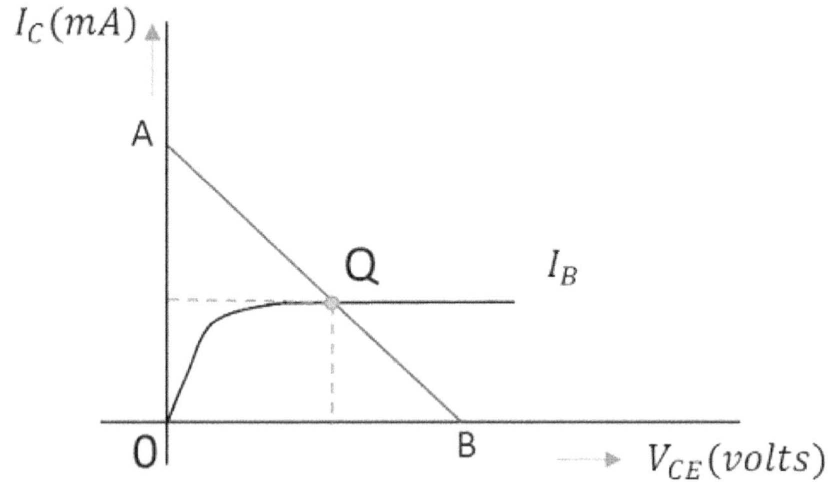

Figure : 2.18 Q-point on a characteristics curve

The load line has to be drawn in order to obtain the Q-point. A transistor acts as a good amplifier when it is in active region and when it is made to operate at Q-point, faithful amplification is achieved.

DC Load Line

When a transistor is biased and no signal is given at its input, load line drawn at such condition can imply the DC condition. Here there will be no amplification as the signal is absent. Following circuit describe such a condition;

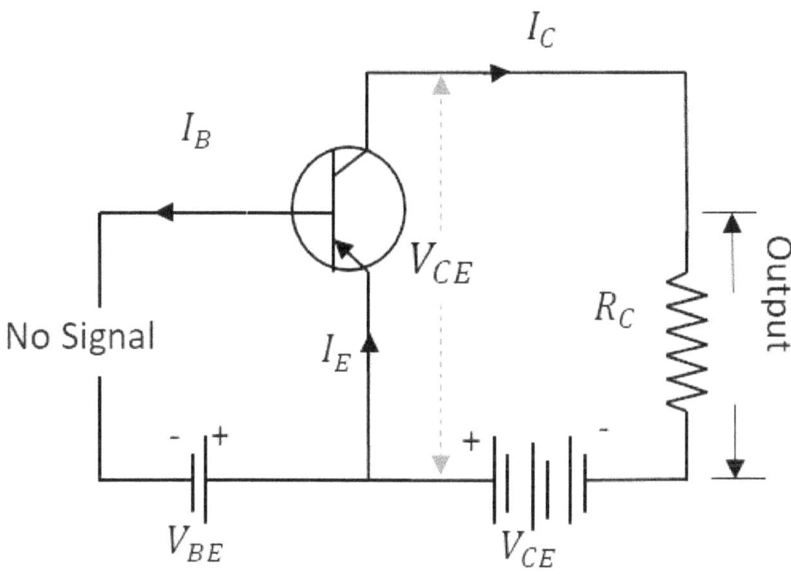

Figure : 2.19 A typical transistor under no signal condition

The value for collector-emitter voltage V_{CE} any given time will be

$$V_{CE} = V_{CC} - I_C R_C$$

As V_{CC} and R_C are fixed values, the above equation is a first degree equation and therefore will be a straight line on the output characteristics.

This line is called as **D.C. Load line**. The figure below shows the DC load line.

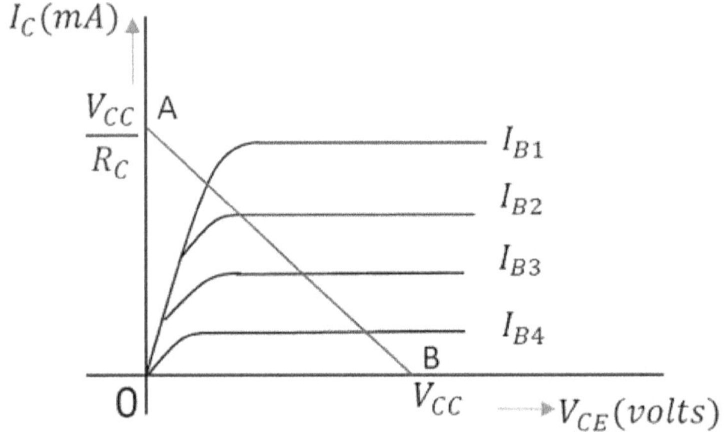

Figure: 2.20 DC load line on an output characteristics curve

Determination of a load line

In order to obtain a load line, the two end points of the straight line are to be determined. Let these two points to be A and B.

To obtain A and B

When the collector-emitter voltage V_{CE} = 0, the collector current is maximum and is equal to V_{CC}/R_C. This gives the maximum value of V_{CE}. This is given as,

$$V_{CE} = V_{CC} - I_C R_C$$

$$0 = V_{CC} - I_C R_C$$

$$I_C = V_{CC} / R_C$$

This gives the point A ($OA = V_{CC}/R_C$) on collector current axis, shown in the above figure.

Similarly, when the collector current I_C = 0, then collector emitter voltage is maximum and will be equal to the VCC. This gives the maximum value of I_C. This is shown as

$$V_{CE} = V_{CC} - I_C R_C$$

$$= V_{CC}$$

(As $I_C = 0$)

This gives the point B, which means (OB = V_{CC}) on the collector emitter voltage axis shown in the above figure.

Hence, we got both saturation point and cut-off. So, a DC load line can be drawn.

SUMMARY

- Being regarded as one of greatest invention of 20th century, transistors became the founding chapter for modern electronics.

- A simple Bipolar Junction Transistors basically made of two PN-junction diodes thereby forming three connecting terminals. Each terminal is specifically named according to their respective properties. These three terminals are known as the Emitter (E), the Base (B) and the Collector (C) respectively.

- Bias is "a potential applied to a device (such as a transistor, diode etc.) to establish a reference point or level for operation".

- When p-type region of a junction is connected to the positive end and n-type region is connected to the negative end of a battery (voltage source), then the junction is said to be forward biased.

- A P-N junction is said to be reverse biased, when the n-type region is connected to the positive end and p-type region is connected to the negative end of the battery (voltage source).

- In Common Base (CB) or grounded base configuration, the base terminal is common to both the input and the output signal.

- α is defined as ratio of change in the collector current from cut-off to emitter current change from cut-off. α is called large signal current gain of transistor in CB configuration.

- Mainly, the transistor circuits that we normally take into use are of common emitter configuration. It is also sometimes called grounded emitter configuration because the emitter is common to input and output both.

- In CB configuration, I_C and I_B are related to each other by a quantity called β.

- Load line is the locus of operating point of a transistor. It is described by a graph plotted between I_C and V_{CE} for any transistor configuration. It helps to determine respective values of collector current and collector to emitter voltage to operate transistor in any particular mode or region.

- The maximum possible value for the collector current can be defined by a point present on the Y-axis. This is known as saturation point.

- The maximum possible value for collector-emitter voltage can be defined by a point on the x-axis. That point is called cut-off point

SHORT ANSWER TYPE QUESTION

Q.1. If the emitter and the base of a transistor have same doping concentration, how will the base-current and the collector-current be affected?

Q.2. Can two p-n junction diodes placed back to back work as a p-n-p transistor? Give reasons to your answer.

Q.3. Under normal use of transistors, the emitter is forward-biased and the collector is réverse-biased. Can either of these biasing be changed? Explain.

Q.4. Which one of the transistors p-n-p and n-p-n is more useful and why?

Q.5. A transistor is a current-operated device while a triode valve is a voltage-operated device. Explain this.

Chapter 3

Field Effect Transistors

A Brief Chapter Overview

FIELD EFFECT TRANSISTORS

3.1 Introduction

Field effect transistors are a range of semiconductor devices that are used to switch, amplify and development of modern integrated chips. Field effect transistors are often shortly called FETs are voltage controlled devices unlike the Bipolar Junction Transistors (BJTs) that are current controlled devices. In BJTs small change in emitter current causes large change in collector current and base current. However there are some limitations in current controlled devices. They always require some current for its proper working and so have higher heating effects. They also consume more power. In comparison to FETs they have low life. So in recent developments in digital and analog electronics FETs are often used.

FETs also have three terminals like BJTs. BJTs have name of emitter (E), base (B) and collector terminal (C). FETs have a separate name of Source (S), Gate (G) and Drain (D). FETs are unipolar devices unlike BJTs. If channel in which electrons flow are n-type, then it is called n-channel FET and if channel is of p-type, then FET is called p-channel FET. Gate voltage acts as controller of current between Source and drain.

3.2 Differences between FETs and BJTs

Some major differences in BJTs and FETs are:

BJT	FET
These are current controlled devices and small change in emitter current causes large change in base and emitter current	These are voltage controlled devices and require current only in turning on the transistor. Small change in Gate voltage cause change in current between Source and drain.
It is a bipolar device. This means it can be operated by both holes and electrons	It is a unipolar device and can be operated by either holes or electrons. One FET cannot have both the charge carriers.

BJT	**FET**
Bipolar junction transistor consist of three terminals namely emitter, base and collector. These terminals are denoted by E, B and C.	Field effect transistor consist of three terminals namely source, drain and gate. These terminals are denoted by S, D and G.
A BJT needs a small amount of current to switch on the transistor. The heat dissipated on bipolar stops the total number of transistors that can be fabricated on the chip.	Only require current to switch on the transistor.
Has low to medium input impedance	Have high input impedance, lesser noise and good amplification at higher frequencies.
Has higher heating effects	Low heating effects
Consumes more power in on state	Low power consumption
Are greater in size than FET	Are smaller in size of same rating

Most common types of FETs

1. Junction Field effect transistors (JFET): The JFET (junction field-effect transistor) uses p–n junction of reverse biased type so the gate and the body are separate.

2. Metal Oxide semiconductor Field effect transistor (MOSFET): MOSFET also called as Insulated Gate Field Effect Transistor (IGFET) utilizes an insulator (like SiO_2) between the gate and the body.

3. Insulated Gate Bipolar Transistor (IGBT): It is a device mainly used for power control. It has a structure similar to a MOSFET coupled with a bipolar-like main conduction channel. These are commonly used for the high voltage i.e., above 200V drain-to-source potential range of operation. However Power MOSFETs are still the device of choice for drain-to-source voltages of small order less than 200 V.

4. Biologically sensitive FET (BioFet): The BioFET is a class of sensors/biosensors based on ISFET (Ion sensitive field effect

transistor) technology which are utilized to detect charged molecules; when a charged molecule is near them, changes in the electrostatic field at the BioFET surface result in a measurable change in current through the transistor. Few of them are EnFETs, DNAFETs, CPFETs, BeetleFETs, and FETs based on ion-channels/protein binding.

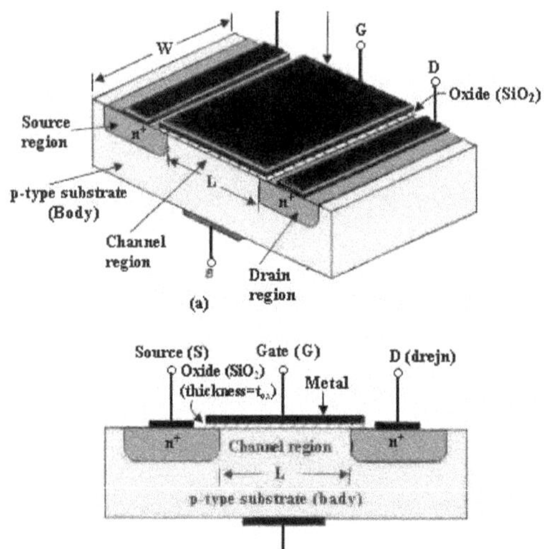

5. Metal Semiconductor FET (MESFET): MESFET substitutes the p–n junction of the JFET with a Schottky barrier. They are used in GaAs and also other III-V semiconductor materials.

6. Fin FET: **FinFET**, also known as Fin Field-Effect Transistor, is a type of non-planar or "3D" transistor used in the invention of modern processors mainly mobile processors. It is built on an SOI (silicon on insulator) substrate. Here gate is kept on two, three, or four sides of the channel or wrapped around the channel, forming a double gate structure. These devices are called "finfets" as the source/drain region forms fins on the silicon surface.

Here we will mainly focus on the one that are more basic and incorporates all the knowledge.

3.3 Junction Field Effect Transistor

JFET are of two types' p-channel and n-channel depending on the type of semiconductor used to make channel of JFET. Their symbols have been shown. The N-channel JFET's channel is always doped with donor type impurities that is the flow of current through the channel is -ve (so given the name of N-channel) made of electrons. Whereas P-channel JFET's channel is always doped with acceptor type impurities that is flow of current through the channel is +ve (so given the name P-channel) made of holes. Now N-channel JFET's always have a greater channel conductivity and low resistance than their comparative P-channel types of JFETs, as electrons have a greater mobility through a conductor than in holes. So N-channel types JFET's more energy efficient conductor than compared to the P-channel.

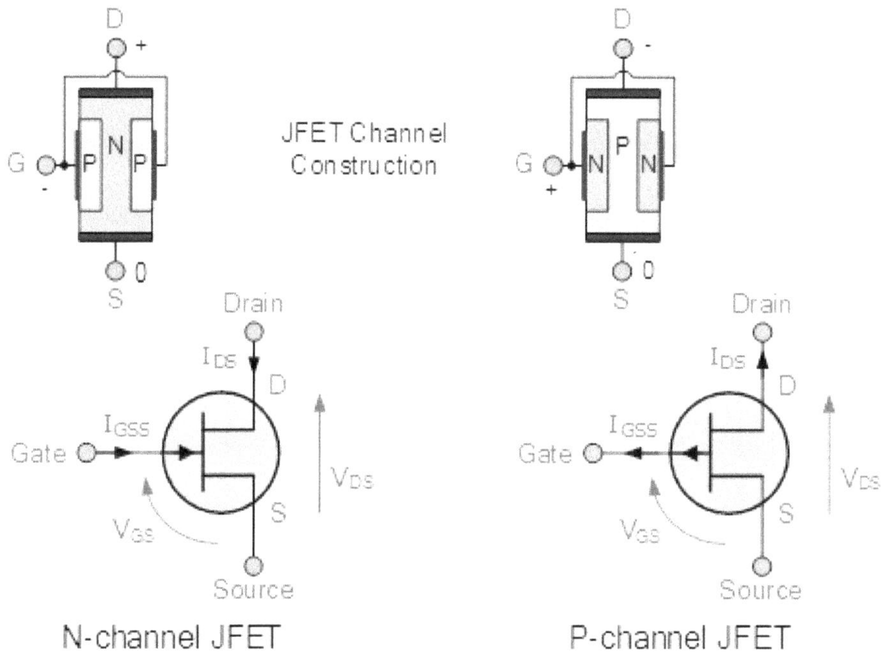

Figure shows that two ohmic connections at either end of the channel called Drain and Source. However within this channel there is a third electrical connection which is called the Gate and this can also be of P-type material or N-type material so there is formation of PN-junction with the main channel.

The semiconductor channel of **Junction Field Effect Transistor (JFET)** is a resistive path inside which the current I_D flows driven by voltage V_{DS} and current is conducted in both the directions. Also as this channel is resistive and so develops voltage gradient through the length of the channel with this voltage becoming less +ve as we start to move from the Drain terminal to the Source terminal.

The result is that the PN-junction therefore has a high reverse bias at the Drain and a lower reverse bias at the Source terminal. This bias causes a depletion layer to be formed within the channel and whose width changes with the bias voltage.

The magnitude of the current flowing inside the channel between the Drain and the Source terminals is controlled by a voltage applied at the Gate terminal, which is reverse-biased voltage. In N-channel JFET this Gate voltage is -ve while for a P-channel JFET the Gate voltage is +ve.

The main difference between the Junction FET and a BJT is that when the JFET junction is reverse-biased the Gate current becomes zero, whereas the Base current of the BJT is always some value above zero.

Working of JFET

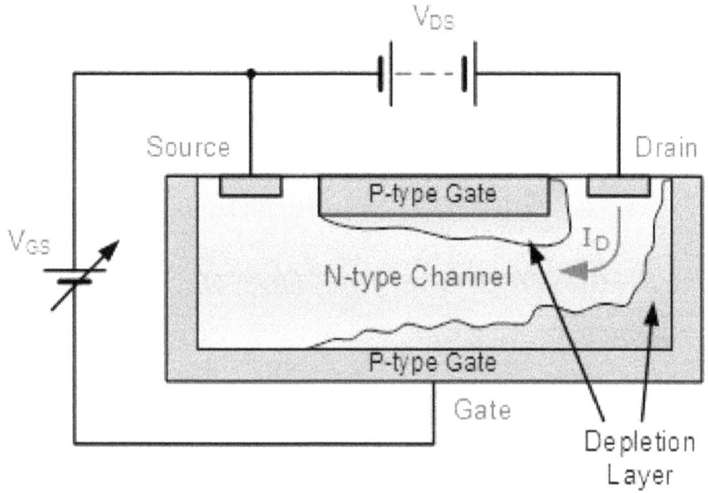

Figure shows an N-type channel with a P-type semiconductor region known as the Gate diffused into the N-type semiconductor channel forming a PN-junction in reverse bias and give rise to formation of *depletion region* around the Gate portion when no external voltage is present. So this is reason why JFETs are known as depletion mode devices.

Depletion layer forms a potential gradient of varying thickness around the PN-junction which restricts the current which is flowing through the channel and in this way increasing the overall electrical resistance of the channel. This is done by changing the effective width of depletion region.

Then we can see the most depleted portion of the depletion layer is in between the drain and gate from the figure, while the least depleted portion is between the Gate terminal and the Source. There the depletion region is very narrow so JFET's channel conducts current with zero bias voltage applied.

When no external Gate voltage (V_G = 0), and a small voltage between drain-source (V_{DS}) applied, maximum saturation current (I_{DSS}) will flow in the channel from the Drain to the Source as there is no restriction from small depletion region.

If a small negative voltage ($-V_{GS}$) is given at the Gate terminal the size of the depletion region starts to further reducing the overall effective width of the channel and so provides resistance to current. So applying reverse bias voltage increases the size of the depletion region reducing the conduction of the channel.

Now as the PN-junction is reverse biased, small current will flow into the gate connection. If the Gate voltage ($-V_{GS}$) is made more -ve, the width of the channel decreases until current between the Drain and the Source is no more and the FET is said to be in pinched-off state (like the cut-off region of Bipolar Junction Transistor). The voltage is known as the "pinch-off voltage", (V_P).

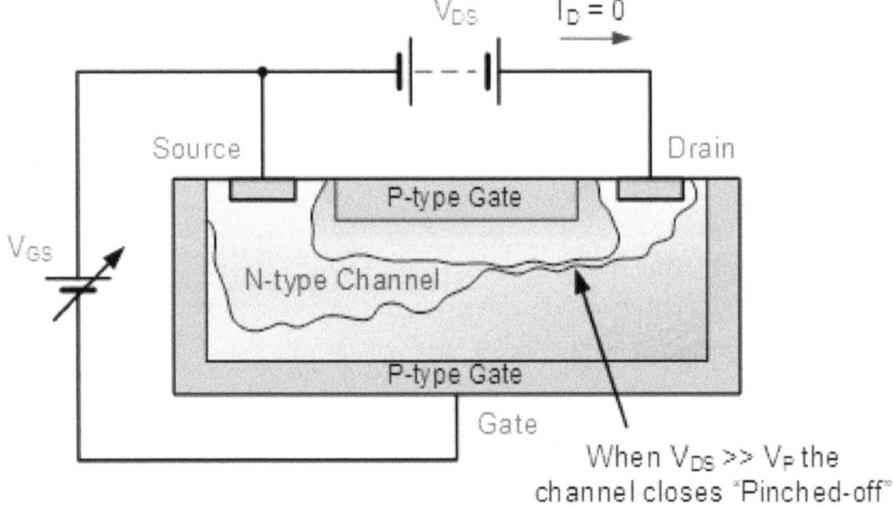

In this pinched-off region the Gate voltage, V_{GS} controls the channel current while V_{DS} plays no role.

This means that the JFET acts more like a voltage controlled device having zero resistance when V_{GS} = 0 and maximum resistance (R_{DS}) when the Gate voltage is most -ve. In normal operating conditions, the JFET always has reverse biased gate-source.

It is essential that the Gate voltage is never +ve then the channel current will not flow to the source, the result will be damage to the JFET. Then to close the channel:

- Gate Voltage (V_{GS}) and V_{DS} should not increased from zero.

- V_{DS} and Gate control should not decreased negatively from zero.

- V_{DS} and V_{GS} keep varying.

The P-channel **Junction Field Effect Transistor** operates exactly the same way as the N-channel, with the following noting things: 1). Channel current is +ve due to presence of holes, 2). Biasing voltage polarity should be in opposite direction

The output characteristics of an N-channel JFET with the gate short-circuited to the source is given as:

The voltage V_{GS} at the Gate controls the current flowing between the Drain and the Source terminals. V_{GS} is the voltage of the Gate terminal with respect to source terminal while V_{DS} refers to the voltage applied between the Drain and the Source.

As a **Junction Field Effect Transistor (JFET)** is a voltage controlled device, so if there is no current flowing into the gate terminal then the Source current (I_S) must be equal to drain current (I_D).

The characteristics curves example shown in the figure shows the four different regions of operation for a JFET and these are:

- Ohmic Region – When V_{GS} = 0 this occurs when depletion layer of the channel is narrow and then JFET acts as a voltage controlled resistor.

- Pinch-off Region – This is also called as the cut-off region where the V_{GS} is sufficient to make the JFET to act as an open circuit as the channel resistance is maximum.

- Saturation or Active Region – The JFET behaves as a good conductor and is controlled by the Gate to Source voltage, (V_{GS}) while the Drain to Source voltage, (V_{DS}) has little or no effect.

- Breakdown Region – The voltage of the Drain with respect to Source, (V_{DS}) is high enough to make the JFET to have a break down and this would lead to pass uncontrolled maximum current.

The characteristics curves for a P-channel JFET will be the same as those above, except that the Drain current I_D will decrease with application of increasing positive Gate-Source voltage, V_{GS}.

The Drain current is zero at point where V_{GS} = V_P. For normal operation, V_{GS} is biased to be at point between V_P and 0. Then we can find I_D for any given bias point in the saturation or active region as follows:

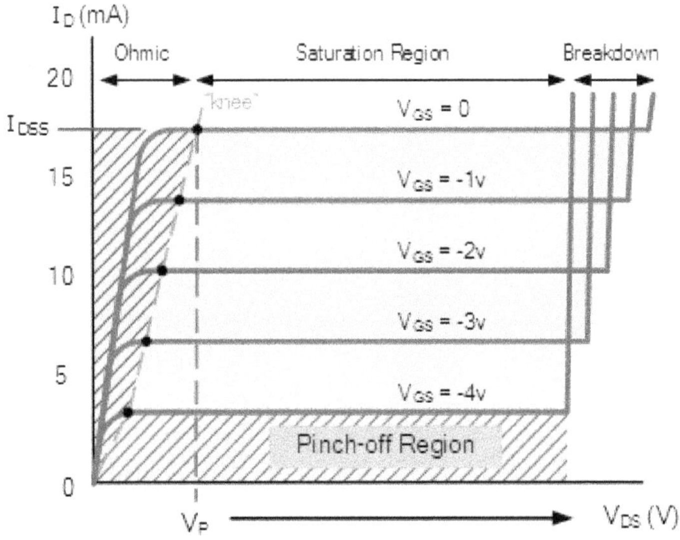

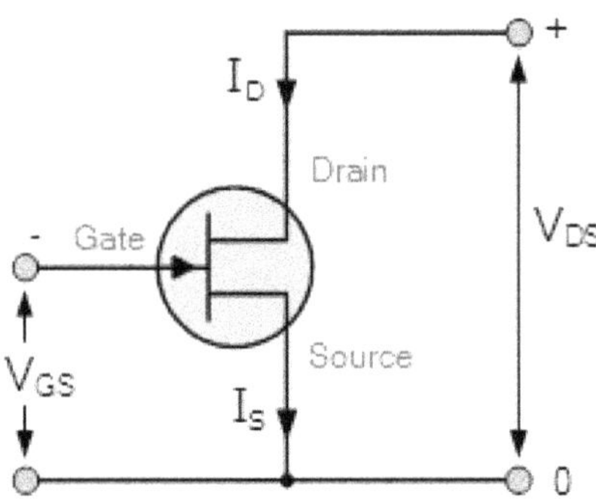

$$I_D = I_{DSS}\left(1 - \frac{V_{GS}}{V_P}\right)^2$$

The above is only valid for active or saturation region. However for Ohmic region this takes form as

$$I_D = I_{DSS}\left[-2\left(1 - \frac{V_{GS}}{V_P}\right)\frac{V_{GS}}{V_P} - \left(\frac{V_{GS}}{V_P}\right)^2\right]$$

At boundary between Ohmic and active region

$$V_{DS} = V_{GS} - V_P$$

So that

$$I_D = I_{DSS}\left(\frac{V_{GS}}{V_P}\right)^2$$

And this is represented by dots in figure i.e., boundary between ohmic and active regions.

Configurations of JFET

1. Common Source Configuration (CS):

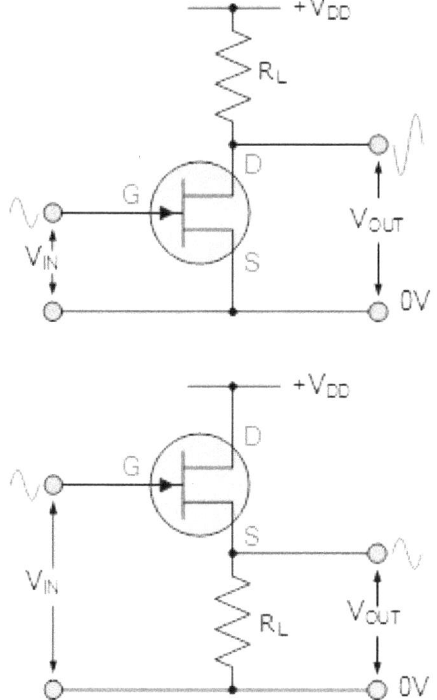

In the **CS** configuration like the common emitter in BJT, the input is applied to the Gate terminal and its output is taken from the Drain as shown in figure. This is the most common mode of operation of the FET due to its high input impedance and good voltage amplification and as such Common Source amplifiers are widely used.

The CS mode of JFET connection is mainly used in audio frequency amplifiers and in high input impedance pre-amplifiers and stages. Being an amplifier, the output signal is 180° "out-of-phase" with the input.

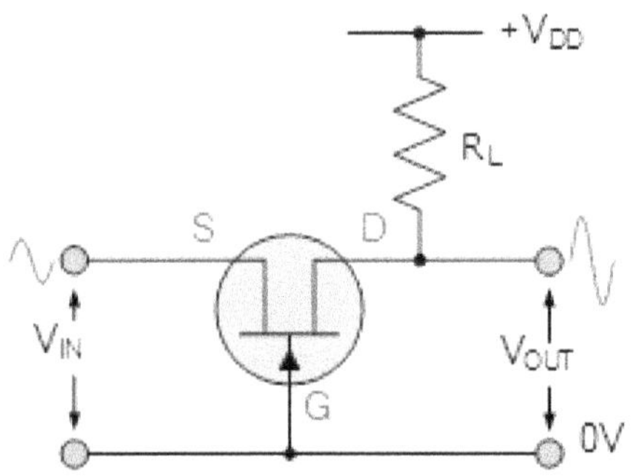

2. Common Gate Configuration (CG):

In the **CG** configuration like the common base of BJT, the input is applied to the Source and its output is taken from the Drain with the Gate connected directly to ground as shown in figure. This configuration has got high impedance feature lost as the common gate has a low input impedance, but high output impedance.

This type of FET configuration are mainly employed in high frequency circuits or in impedance matching circuits where a low input impedance needs to be matched to a high output impedance. Here output is "in-phase" with the input.

3. Common Drain Configuration (CD):

In the **CD** configuration like the common collector of BJT, the input is applied to the Gate terminal and its output is taken from the Source terminal. The CD or "source follower" configuration has very high input impedance and low output impedance and near-unity voltage gain so is mainly employed in buffer amplifiers. The voltage gain of this configuration is little less than unity and the output signal is "in-phase", 0° with respect to input signal.

This configuration is often referred to as "Common Drain" as there is no signal at the drain connection, the voltage present, $+V_{DD}$ just provide a bias. The output is in-phase with the input voltage.

Small Signal Model of JFET

This model is for low frequency and small signal operation of transistor. The relationship between various parameters are given as

$$g_m = transconductance = \frac{I_D}{V_{GS}}$$

$$g_m = g_{mo} \sqrt{\frac{I_D}{I_{DSS}}}$$

Where g_{mo} is transconductance for $V_{GS} = 0$ and is $-2\frac{I_{DSS}}{V_P}$

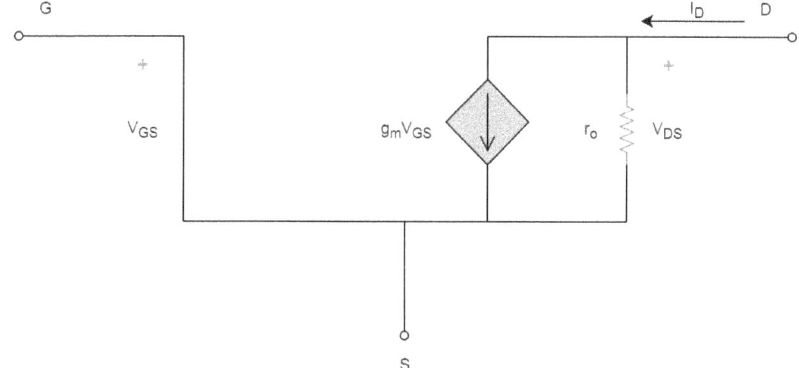

As well **because the** Junction Field **result electronic transistor** (JFET), **there's** another **sort of** Field **result electronic transistor on the market** whose Gate input is electrically insulated from **the most** current carrying channel and is **thus known as associate** Insulated Gate Field **result electronic transistor**

The most widely recognized kind of protected door FET which is utilized in a wide range of sorts of electronic circuits is known as the Metal Oxide Semiconductor Field Effect Transistor (MOSFET). It is a voltage controlled field impact transistor that contrasts from a JFET in that it has a "Metal Oxide" Gate cathode which is electrically protected from the principle semiconductor n-channel or p-channel by an extremely meager layer of protecting material as a rule silicon dioxide

- Depletion Type – the transistor requires the Gate-Source voltage, (V_{GS}) to switch the gadget "OFF". The exhaustion mode MOSFET is equal to a "Regularly Closed" switch.

- Enhancement Type – the transistor requires a Gate-Source voltage, (V_{GS}) to switch the gadget "ON". The upgrade mode MOSFET is identical to an "Ordinarily Open" switch.

3.4 Depletion Type Mosfet

Basic construction

The essential development of the n-channel consumption type MOSFET is given in fig 3.1. a slab of p type material is framed from a silicon base and is alluded to as the substrate. it is the establishment whereupon the gadget will be constructed. in a few cases the substrate is inside associated with the source material. in any case, numerous discrete gadget give an extra terminal name SS bringing about a four terminal gadget, for example, that showing up in fig. 3.2. the channel and the source terminal are connected through metallic contact to n-doped locale connected by a n direct as appeared in the fig. the entryway is additionally associated with a metal contact surface yet remain protected from the n-channel by a exceptionally slight silicon dioxide(sio$_2$)layer. Silicon dioxide is a specific sort of insulator referred as a dielectric when exposed to an externally applied field.

The way that the silicon dioxide layer is a protecting layer uncover the accompanying reality.

There is no a direct electrical connection between the gate terminal and the channel; of MOSFET

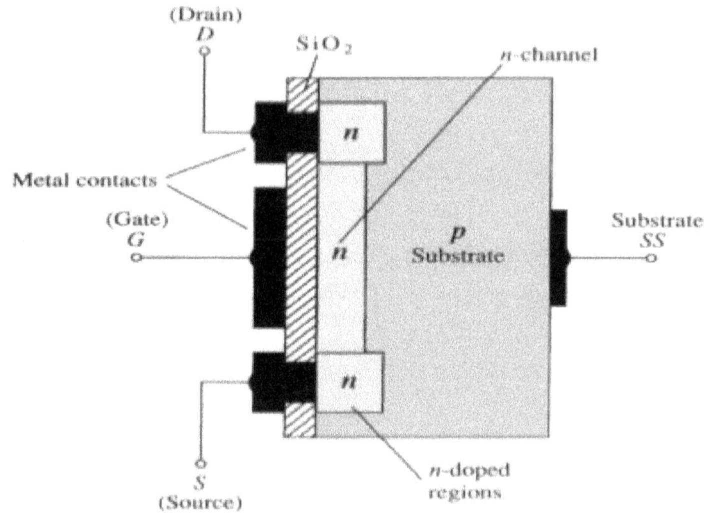

Fig 3.1 n-channel depletion type MOSFET

THE INPUT RESISTANCE OF A MOSFET IS OFTEN THAT OF THE TYPICAL JFET,EVEN THOUGH THE INPUT IMPEDANCE OF MOST JFET IS SUFFICIENTLY HIGH FOR MOST APPLICATION. THE VERY HIGH INPUT IMPEDANCE CONTINUE TO FULLY SUPPORT THE FACT THAT THE GATE CURRENT (IG) is essentially zero ampere for DC BIASED Configuration.

The purpose behind the metal oxide semiconductor FETis now genuinely evident metal for the channel, source and gate association with the best possible surface specifically the gate terminal and the control to be offered by the surface territory of the contact the oxide for the silicon dioxide protecting layer and the semiconductor for the essential structure on which the n-type and f p-type area are diffused. The protecting(insulating) layer between the door and direct has brought about another name for the gadget; protected gate FET, despite the fact that this name is utilized less and less in current writing.

3.5 Basic Operation and Characterisation

The gate to source voltage in fig 3.3 is set to zero volts by the immediate association from one terminal to the terminal and a voltage V_{DS} is applied over the drain to source terminal. The result is collaboration for the positive terminal at the channel by the free electron of the n-channel and the current like that set up through the channel of the FET. Actually the subsequent (resulting) current with VGS is equivalent to zero volt keep on being marked I_{DSS} as appeared in fig. 3.4

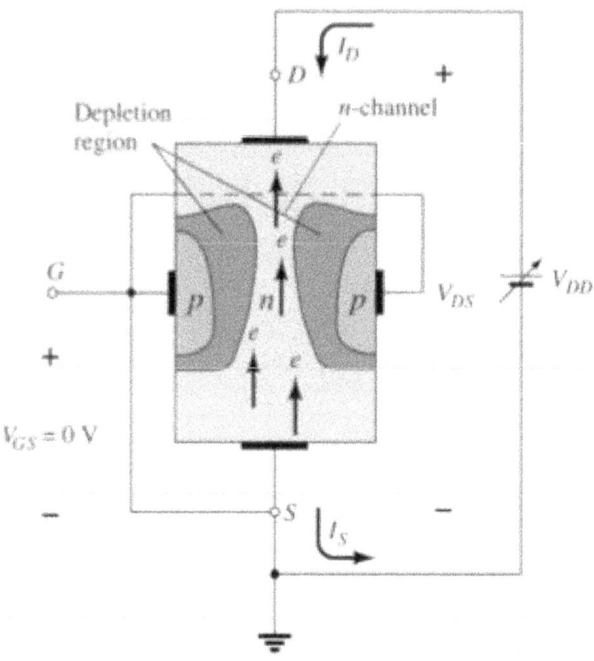

Fig 3.3 n-channel depletion type MOSFET with vgs=0volt and vdd applied voltage

In fig 3.4, V_{GS} has been set a negative voltage like -1 potential unit. The negative potential at the gate can tend to pressure electron toward the p-type substrate and attract holes from the p-type substrate as shown in fig 3.5.depending on the magnitude of the negative bias established by V_{GS} grade of recombination between electron and holes will occur which will decrease the amount of electron within the n-channel for conduction. The additional negative the bias the upper the speed of recombination. The net level of drain current is thus reduced with increasing negative bias for VGS as shown in fig one.4 for V_{GS}=-1 potential unit,-2 potential unit then on, to the pinch off level of -6volt. The ensuing level of drain current and therefore the plotting of the transfer curve issue precisely as delineate for the field-effect transistor.

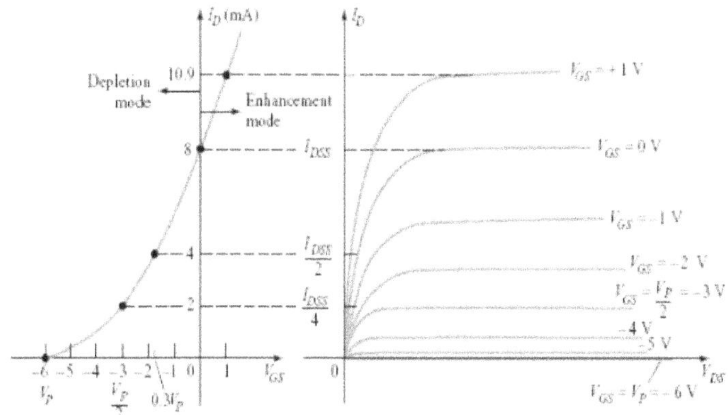

Fig 3.4 Characteristics for an n-channel depletion t

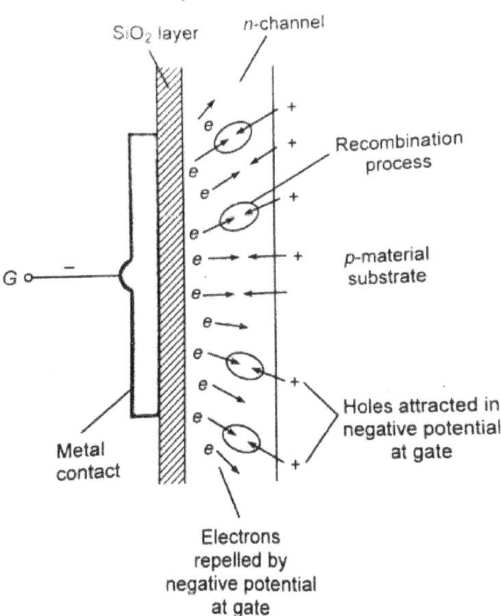

For the positive value of V_{GS}, the positive gate can draw further electron from the p-type substrate due to the reverse escape current and establish new carrier through the collision ensuing between fast particles. as the gate to supply voltage still increase within the positive direction. fig 3.5 reveal that the drain current can increase at a speedy rate for the explanation listed higher than. The vertical spacing between the VGS is up to zero V and VGs is equal to one volt curve of fig 3.6 is evident indication of what quantity this has inflated for the 1-v channel in VGS. thanks to the speedy rise, the user should bear in mind of the utmost drain current rating since it can be exceeded with a positive gate voltage. that is for the device of fig 3.6 the applying of voltage VGS is up to four volt would end in drain current of 22.2 mA, that may presumably exceed the great rating for the device. As disclosed higher than, the application of a positive gate to supply Voltage has increased the amount of free carriers within the channel compared to it encountered with VGS up to zero volt For this reason the region of positive gate voltage on the drain or transfer characteristics is commonly mentioned because the sweetening region. with the region between bring to an end and therefore the saturation level of I_{DSS} mentioned because the depletion region.

It is notably fascinating and useful that William Bradford Shockley equation can still be applicable for the depletion kind MOSFET characteristics in each the depletion and sweetening region. for each region, it is merely necessary that the right sign be enclosed with VGS within the equation and therefore the sign rigorously monitored in the same mathematical operations.

p-channel depletion type MOSFET

The construction of p-channel depletion sort MOSFET is exactly the reverse of that showing in fig 3.1. That is, there is currently Associate in Nursing n-type substrate and p-type channel as shown in on top of fig 3.9(a). The terminal stay as known however all the voltage polarities and therefore the current direction are reversed as shown within the figure. The drain characteristics would seem exactly in fig 3.4 however with V_{DS} having negative values as indicated and V_{GS} having the opposite polarity as shown in fig. The reversal in V_{GS} can lead to a relaxation(mirror image)on for the transfer characteristics as shown in fig 3.9(b). In different word the drain current can increase from cut-off at $V_{GS}=V_P$ within the positive V_{GS} region to I_{DSS} and so still increase for increasingly negative values of v_{GS} shockley's equation continues to be applicable and needs merely putting the right sign for each V_{GS} and V_P in the equation shockley's condition is as yet pertinent and requires essentially setting the right sign for both V_{GS} and V_P in the equation.

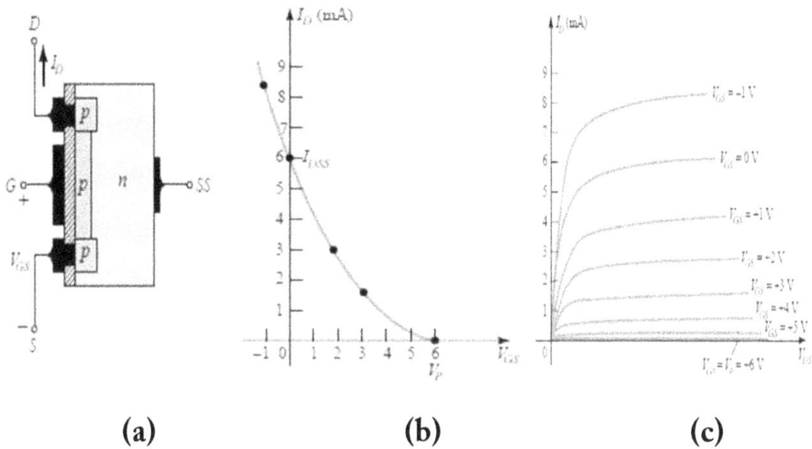

(a)   **(b)**   **(c)**

Fig 3.5 p-channel depletion type MOSFET with I_{DSS}=6mA and v_p=6v

3.6 Enhancement type MOSFET

In spite of the fact that there are a few resemblances in development and mode of operation between depletion type and enhancement type MOSFET, the characteristics of the enhancement type MOSFET are very unique in relation to anything acquired so far. The exchange curve isn't characterized by shockley's equation and the drain current is presently sliced off until the gate to source voltage achieves a particular size. specifically current control in an n-channel device is currently affected by a positive gate to source voltage instead of the scope of negative voltages experienced for n-channel JFET and n-channel depletion type MOSFETs

Basic construction

The fundamental development of the n-channel enhancement type MOSFET is given in fig 3.6.A section of p-type material is formed from a silicon base and is again mention to as the substrate. similarly as with the depletion type MOSFET, the substrate is at some point inside associated with the source terminal, while in other case a fourth lead is made available for outside control of its potential dimension. The source and the drain terminal are again associated through metallic contact to n-doped region, but note in fig 3.6 the absence of a channel between the two n-doped regions. This is the essential distinction between the development of depletion type and enhancement type MOSFETs—the absence of the channel as a built part of the device. The sio2 layer is as yet present to isolate the gate metallic stage from the area between the drain and the source, however at this point it is basically isolated from a segment of the p-type material. in summary, therefore the development of the enhancement type MOSFET is very like that of the depletion type MOSFET is very like that of the depletion type MOSFET aside from the nonappearance of a channel between the drain and the source terminals.

Basic Operation and Characteristics

When V_{GS} is set 0 volt and a voltage connected between the drain and source of the device of fig 3.6, the absence of a n-channel will result in a current of successfully zero amperes-very unique in relation to the depletion type MOSFET and JFET where ID=IDSS. It isn't sufficient to have a huge accumulation of carrier (electron) at the drain and source. If way neglect to exist between the two. With VDS some positive voltage, VGS at 0 volt and terminal ss directly associated with the source there are in actuality two reverse biased p-n intersection between the n-doped area and the p-substrate to contradict any significant stream among drain and source.

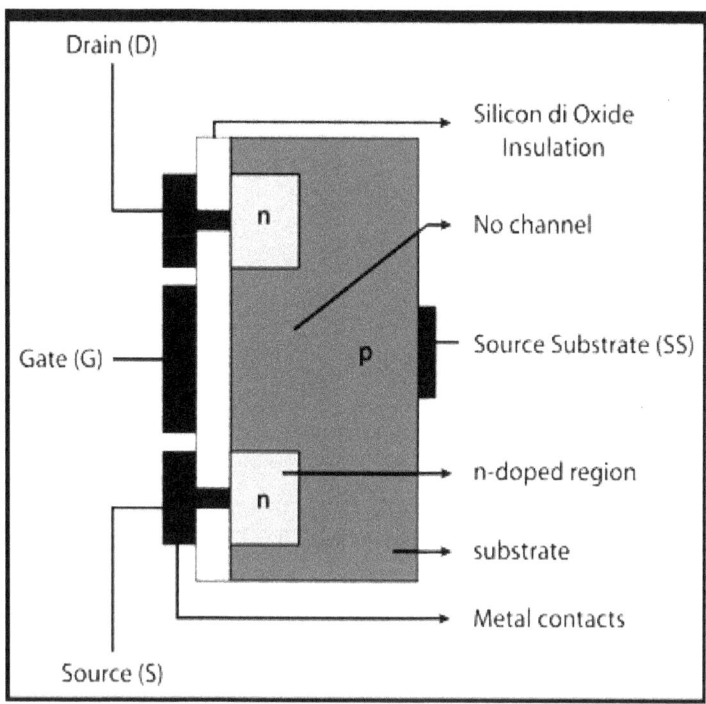

Fig 3.6 n-channel enhancement type MOSFET

In fig 3.7 the two vDS and vGS have been set at positive voltage more prominent than 0V. Setting up the channel and gate at a positive potential regarding the source. The positive potential at the gate will pressure the holes in p-substrate along the edge of the Sio2 layer to leave the area and enter further locale of the substrate, as appeared in fig...the result in consumption district close to the Sio2 protecting layer bereft of holes. however, the electron in the p-substrate will be pulled in to the positive door and collect in the area close to the outside of the Sio2 layer. The sio2 layer and its protecting characteristics will keep the negative bearers from being assimilated at the gate terminal.

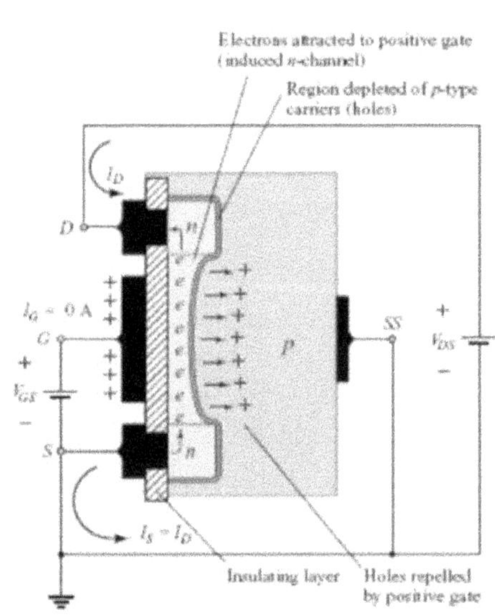

Fig 3.7 channel formation in the n-channel enhancement type MOSFET

As VGS increase in magnitude, the concentration of electron near the Sio2 surface increase until eventually the induced n-type region can support a measurable flow between drain and source. the level of VGS that result in the significant increase in drain current is called threshold voltage and is given by the symbol VT.on specification sheet it is referred to as VGS (The),although VT is less unwieldy and will be used in the analysis to follow. since the channel is nonexistent with VGS =0volt an enhanced by the application of positive gate to source voltage, his type of MOSFET is called an enhancement –type MOSFET. both depletion and enhancement type MOSFET have enhancement type region but the label was applied to the latter since it is its only mode of operation.

As V_{GS} extend in magnitude, the gathering of electron close to the Sio_2 surface expand until ultimately the brought on n-type area can assist a measurable drift between drain and source. the level of V_{GS} that result in the consequential increment in drain contemporary is referred to as threshold voltage and is given by using the symbol (V_T). on specification sheet it is referred to as V_{GS} (T_h),although V_T is much less unwieldy and will be used in the evaluation to follow. in view that the channel is nonexistent with V_{GS} =0volt and enhanced via the utility of positive gate to source voltage, this type of MOSFET is referred to as an enhancement –type MOSFET. Both depletion and enhancement type MOSFET have enhancement type location however the label was utilized to the latter considering the fact that it is its only mode of operation.

As V_{GS} is accelerated beyond the threshold level, the density of free carrier in the precipitated channel will increase, resulting in an increase degree of drain modern. however, if we hold V_{GS} consistent and enlarge the value of V_{DS}, the drain current will finally reach a saturation stage as happened for the JFET and depletion type MOSFET. The levelling off of I_D is due to a pinching off process depicted by way of the narrower channel at the drain give up of the triggered channel as shown in fig 3.8. applying Kirchhoff voltage law to the terminal voltage of the MOSFET of fig 3.8, we find that

$$V_{DG} = V_{DS} - V_{GS} \qquad \qquad ...(3.1)$$

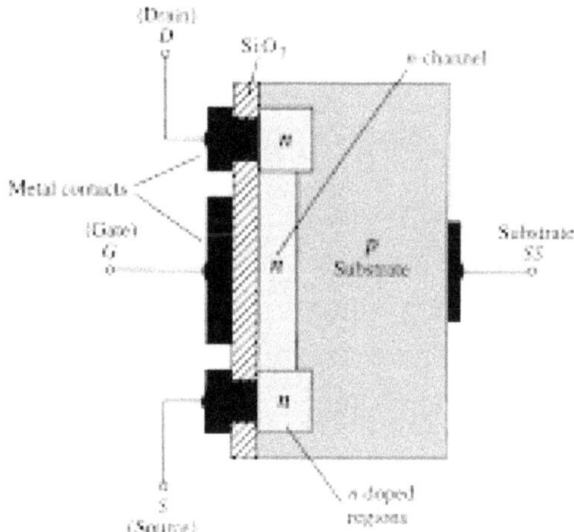

Fig 3.8 change in channel and depletion region with increasing level of V_{DS} for a fixed value of V_{GS}

If V_{GS} is held fixed at some value such as 8V and VDS is increased from 2 to 5 V, the voltage V_{DG}(by eq 3.1) will drop fig from -6 to -3 V and the gate will become less and less positive with respect to then drain. This decrease in gate to drain voltage will in turn diminish the appealing powers with the expectation of complimentary bearers (electron) in this region of the initiated channel, causing a decrease in the effective channel width. Eventually, the channel will be decreased to the point of pinch - off and an immersion condition will be set up as explained before for the JFET and the depletion– type MOSFET. At the end of the day, any further increment in VDS at the fixed estimation of V_{GS} won't influence the saturation level of I_D until breakdown condition are experienced.

The drain characteristics of fig 3.9 reveal that for the device of fig 3.8 with VGS=8v, saturation occurred at a level of VDS=6v.in fact, the saturation level for VDS is related to the level of applied VGS by

$$\mathbf{V_D\,S_{SAT} = V_{GS} - V_T} \tag{3.2}$$

Clearly in this manner for a fixed estimation of VT then the higher the value of VGS the more the saturation level for VDS as appeared in fig 3.8 by the locus of saturation levels. For the characteristics of fig 3.8 the dimension of VT is 2V, as uncovered by the way that the channel current has dropped to 0mA. In general, therefore:

For value of VGS less than the threshold level, the drain current of an enhancement –type MOSAFET is 0 mA

Fig 3.9 clearly reveals that as the level of VGS increased from VT to 8v, the resulting saturation level for ID also increased from a level of 0 to 10 Main addition, It is quite noticeable that the spacing between the level of VGS increased as the magnitude of VGS increased, resulting in ever-increasing increment in drain current.

For levels of VGS>VT, the drain current is related to the applied gate to source voltage by the following nonlinear relationship

$$\mathbf{I_D = K\,(V_{GS} - V_T)^2} \tag{3.3}$$

Again in equation 3.3 is a square term result in the relationship between ID and VGS. Where K is constant and it is the function of the construction of the device. The value K can be determined from the following equation (equation 3.3) where $I_{D(on)}$ and $V_{GS(on)}$ are the values for each at a particular point on the characteristics of the device.

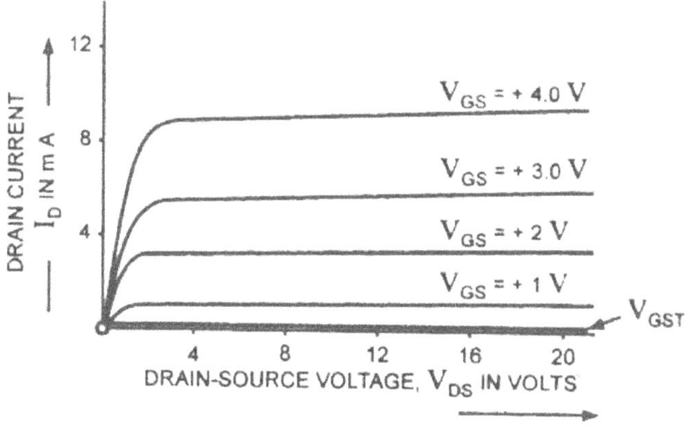

Drain Characteristics

Fig 3.9 Drain characteristics of n channel enhancement type MOSFET

$$K = \frac{I_{D(on)}}{(V_{GS(on)} - V_T)^2} \qquad ...(3.4)$$

substituting ID(on)=10 Ma when VGS(on)=8v from the characteristics of fig 3.9 yields

$$K = \frac{10 Ma}{(8V-2V)^2}$$

$$= \frac{10\ Ma}{(6)^2}$$

$$= \frac{10}{36}$$

$$= 0.278 \times 10^{-3}\ A/V^2$$

And a general equation for ID for the characteristics of fig 3.9 result in:

$$ID = 0.278 \times 10^{-3}\ (V_{GS} - 2V)^2$$

Substituting VGS=4V, we find that

$$ID = 0.278 \times 10^{-3}\ (4V-2V)^2$$

$$= 0.278 \times 10^{-3}\ (2V)^2$$

$$= 0.278 \times 10^{-3} \ (4)$$

$$= \mathbf{1.11ma}$$

As verified by fig. 3.9.at VGS =VT, the squared term is 0 and ID=0ma.

In fig 3.10 the drain and transfer characteristics have been set side by side to describe the transfer process from one to another. Essentially, it proceeds as introduced earlier for the MOSFET. For this situation anyway it must be recalled that the channel current is 0ma for VGS<=VT. At this point a measurable current will result for id and will increment as characterized by Equation (3.3). Note that in defining the point on the transfer characteristics from the drain characteristics, only the saturation level are employed, thereby limiting the region of operation to level of VDS greater than the saturation level are defined by equation (3.2)

The transfer curve fig. 3.10 is certainly quite different from those obtained earlier. For n-channel device, it is currently absolutely in the positive VGS district and does not ascend until VGS=VT. The question presently surface with respect to how to plot the exchange attributes given the dimension of K and VT is incorporated underneath for a specific MOSFET

$$I_D = 0.5 \times 10^{-3} \ (V_{GS} - 4V)^2$$

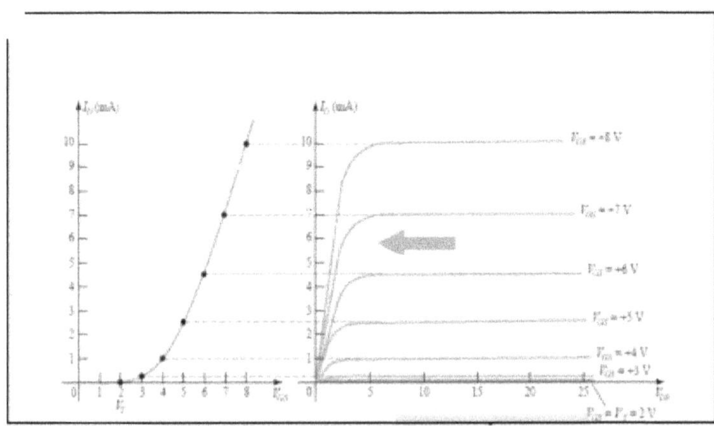

Fig. 3.10 Transfer characteristics for n-channel enhancement type MOSFET from the drain CHARACTERISTICS

Firstly, a horizontal line is draw at ID =omA from VGS=4V as appeared in fig 3.10, next a dimension of VGS more than VT, for example, 5V is picked and substituted into Equation (3.3) to decide the subsequent dimension of ID as pursue

$$I_D = 0.5 \times 10^{-3}(V_{GS} - 4V)^2$$

$$= 0.5 \times 10^{-3}(5V - 4V)^2$$

$$= 0.5 \times 10^{-3}(1)^2$$

$$= 0.5 \text{ mA}$$

p-channel Enhancement Type MOSFETs

The development of a p-channel improvement type MOSFET is actually the switch the of that showing up in fig 3.6 as appeared in fig 3.11(a). that is there now a n-type substrate and p-doped region under the drain (channel) and source association the terminals stay as recognized yet all the voltage polarities and the present bearing are turned around. the channel attributes will show up as appeared in fig. 3.11(c), with increasing level of current coming about because of progressively negative value of VGS. the transfer characteristics will be the perfect representation of the mirror image of fig 3.10,with ID expanding with progressively negative estimation of VGS beyond VT, as appeared in fig 11.11

Equation (3.1) through (3.4s.) are similarly appropriate to p-channel devices.

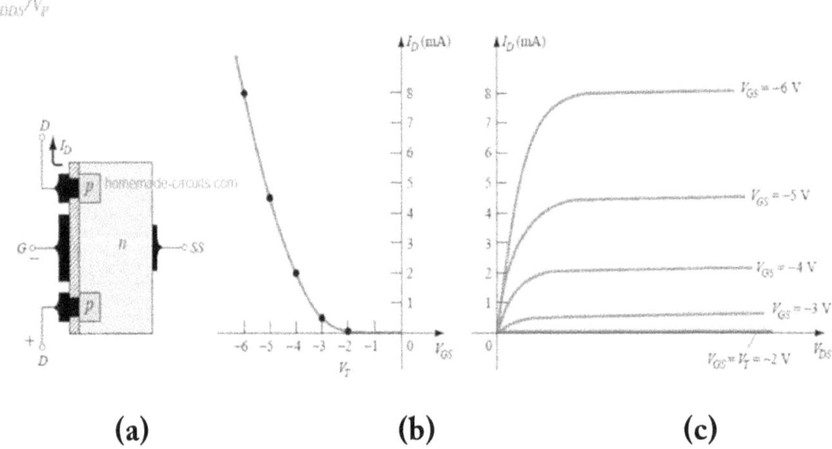

Fig 3.11 *p*-channel enhancement type MOSFET
with VT = 2V and K = 0.5 × 10³ A/V²

3.7 Comparison between Enhancement and Depletion type MOSFET

Enhancement Mode

MOSFETs with improvement modes are often switched on by powering the gate either above the supply voltage for NMOs or below the supply

voltage for the PMOS. In most circuits, this suggests that propulsion a MOSFET gate voltage into the escape boost mode becomes ON. For N-type discharging devices, the edge voltage may be concerning -3 V, therefore it may be stopped by dragging the three V negative gate (leakage by comparison is a lot of positive than the NMOS source). In PMOS, polarities are reversed.

The mode are often determined by the voltage threshold sign (gate voltage versus supply voltage at the purpose wherever solely a layer inversion is made within the channel): for an N-type transistor, modulation devices have positive and depleted thresholds – modulated devices have negative thresholds; for a P-type transistor, positive mode to enhance negative mode, depletion.

Depletion MOSFET

Junction field effect transistors (JFET) are the depletion mode that the door intersection would transmit the bias if the gate was taken in excess of a bit from the source to the channel voltage. Such gadgets are utilized in gallium-arsenide and germanium chips. Likewise note the MOSFET circuit type N exhaust channel image. Because of its development, it offers extremely high section quality (roughly 1010 to 1015). Huge current streams for VDS information at 0 volts VGS.

At the point when the door (i.e, a capacitor plate) is made positive, the channel (i.e., the other capacitor plate) will have a positive charge incited in that. This will prompt the consumption of the significant bearers (i.e, electrons) and therefore the decrease in conductivity.

Cs Amplifier

If we tend to apply little time-varying signal to the input, then under the proper circumstances the mosfet circuit will act as a linear electronic amplifier providing the transistors Q-point is somewhere close to the middle of the saturation region, and therefore the input is small enough for the output to stay linear. Think about the fundamental mosfet electronic amplifier circuit below.

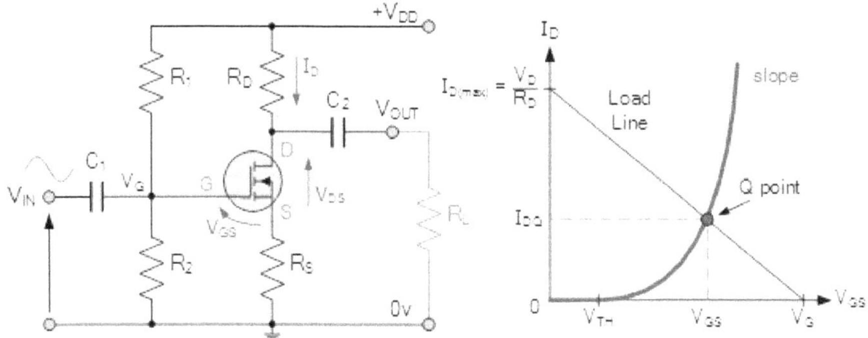

This straightforward enhancement mode common source mosfet amplifier arrangement utilizes a single supply at the channel and creates the required door voltage, VG utilizing a resistor divider.

We recollect that for a MOSFET, no current flow into the gate terminal.

$$V_{DD} = I_D R_S = +V_{DS} + I_D R_S$$

$$= I_D (R_D + R_S) + V_{DS}$$

$$R_D + R_S = V_{DD} - V_{DS}$$

$$I_D$$

So we can write as,

$$R_D = V_{DD} - V_D$$

$$ID$$

$$\text{and } R_S = V_S$$

$$I_D$$

$$V_{GS} = V_G - I_G R_S$$

FOR operation of mosfet, the gate-source voltage must be greater than the threshold voltage, i.e., $V_{GS} > V_{TH}$. Since $I_S = I_D$, the gate voltage, V_G is therefore equal too:

$$VGS = VG - ID\ RS$$

$$VG = VGS + ID\ RS$$

$$VG = VGS + VS$$

To set the mosfet gate voltage to this esteem we select the estimations of the resistors, R1 and R2 inside the voltage divider system to the right qualities. As we probably are aware from over, "no current" flow into the gate terminal of a mosfet device so the equation for voltage division is given as:

The primary objective of a MOSFET amplifier or any amplifier besides is to deliver an output signal that is a dependable proliferation of its input signal however enhanced in extent. This input signal could be a current or a voltage, yet for a mosfet device to work as an amplifier it must be one-sided to work inside its saturation region.

3.8 Mosfet Amplifier Summary

There are two essential kinds of enhancement mode MOSFETs, n-channel and p-channel and in this mosfet amplifier instructional exercise we have taken the n-channel enhancement MOSFET is regularly alluded to as a NMOS, as it very well may be worked with positive gate and drain voltages in respect to the source rather than the p-channel PMOS which is worked with negative gate and channel voltages in respect to the source.

The saturation region of a MOSFET amplifier is its steady present area over its edge voltage, VTH. Once accurately one-sided in the immersion area the channel current, ID shifts because of the gate to-source voltage, VGS and not by the channel to-source voltage, VDS since the drain current is called saturated.

In an upgrade mode MOSFET, the electrostatic field made by the utilization of a gate voltage improves the conductivity of the channel, instead of exhaust the divert as on account of a consumption mode MOSFET.

The edge voltage is the base door predisposition required to empower the development of the channel between the source and the drain. over this esteem the drain current increments in extent to (VGS – VTH)2 in the immersion district enabling it to work as an amplifier

BIASING IN JFET AND MOSFET

Here we will see different methods used to bias JFET and MOSFET. Like with the BJT, the purpose of biasing is to select the proper dc gate-to-source voltage to establish a desired value of drain current and, thus, a proper Q-point. In JFET three types of bias are self-bias, voltage-divider bias or fixed bias, and current-source bias and in MOSFET we see feedback bias and fixed bias for Enhancement MOSFET and self bias and fixed bias for Depletion MOSFET.

JFET BIASING

3.9 Self Bias

As JFET must be operated such that the gate-source junction is always reverse-biased. This condition requires a negative V_{GS} for an n-channel JFET and a positive V_{GS} for a p-channel JFET. So we apply self bias.

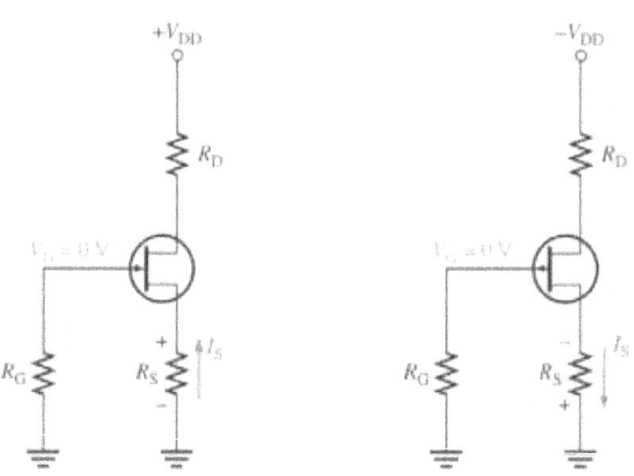

The gate resistor, R_G, does not affect the bias because it has essentially no voltage drop across it; and therefore the gate remains at 0 V. R_G is necessary only to force the gate to be at 0 V and to isolate an ac signal from ground in amplifier applications.

For the n-channel JFET as shown in the figure, I_S produces a voltage drop across R_S and makes the source positive with respect to ground. Since $I_S = I_D$ and $V_G = 0$, then $V_S = I_D R_S$. The gate-to-source voltage is then given by

$$V_{GS} = V_G - V_S = -I_D R_S$$

For p-channel JFET, R_S produces a negative voltage at source making gate at higher potential than source so

$$V_{GS} = I_D R_S$$

Now Drain voltage with respect to ground is determined as

$$V_D = V_{DD} - I_D R_D$$

Also source voltage with respect to ground is determined as $V_S = I_D R_S$

So drain to source voltage is

$$V_{DS} = V_D - V_S = V_{DD} - I_D (R_D + R_S)$$

Q Point for self-biased JFET

To establish a JFET bias point is to determine I_D for a desired value of V_{GS} or vice versa. Then we find the value of R_S using the following relationship. Note That vertical lines indicate an absolute value.

$$R_S = \left| \frac{V_{GS}}{I_D} \right|$$

Now I_D can be found from

$$I_D = I_D SS \left(1 - \frac{V_G S}{V_{GS_{in\ off\ state}}} \right)^2$$

For determining the Q-point of the electric circuit, a self-bias dc-load-line is established on the graph as follows. First, we calculate V_{GS} when $I_D = 0$.

This establishes a point at the origin on the graph. Next, calculate V_{GS} when $I_D = I_{DSS}$.

This establishes second point on graph. Using these points draw load line and where it intersects transfer characteristics curve is Q-point.

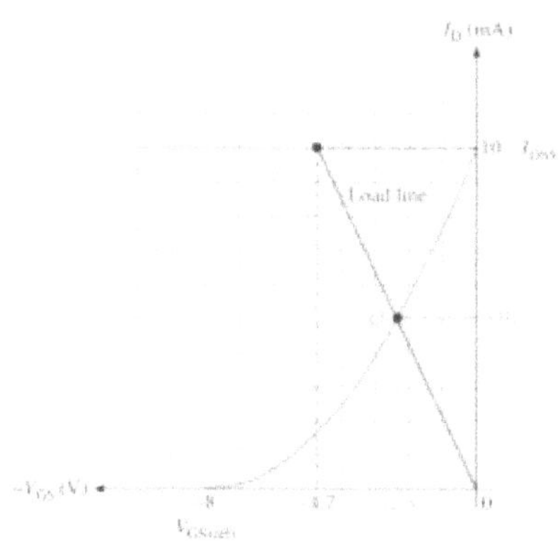

For increased Q-point stability, the value of R_S in the self-bias circuit is increased and connected to a negative supply voltage. This is sometimes called dual-supply bias.

3.10 Voltage Divider Bias

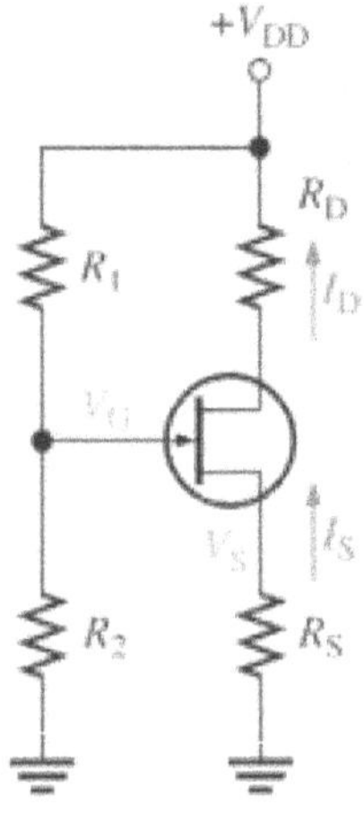

N-channel JFET is shown. Voltage at source of JFET must be more positive than gate voltage so that gate-source junction is reverse biased.

Source voltage is given by

$$V_S = I_D R_S$$

The gate voltage is then given by voltage-divider rule

$$V_G = \left(\frac{R_2}{R_1 + R_2} \right) V_{DD}$$

The gate-source voltage is $V_{GS} = V_G - V_S$

And also drain current is

$$I_D = \left(\frac{V_S}{R_S} \right)$$

Then

$$I_D = \frac{V_G - V_{GS}}{R_S}$$

Finding Q point in Voltage-Divider

An approach similar to the one used for self-bias can be used with voltage-divider bias to graphically determine the Q-point of a circuit on the transfer characteristic curve. In a JFET with voltage-divider bias when $I_D = 0$, V_{GS} is not zero, as in the self-biased case, because the voltage divider produces a voltage at the gate independent of the drain current. The voltage-divider dc load line is determined as follows.

For $I_D = 0$

$V_S = I_D R_S = 0\ V$

$V_{GS} = V_G - V_S = V_G - 0 = V_G$

Therefore, one point on the line is at and $V_{GS} = V_G$

For $V_{GS} = 0$

$$I_D = \frac{V_G - V_{GS}}{R_S} = \frac{V_G}{R_S}$$

A second point on line is at this point. Draw line with these two points and the line that intersects transfer characteristic curve is the Q-point.

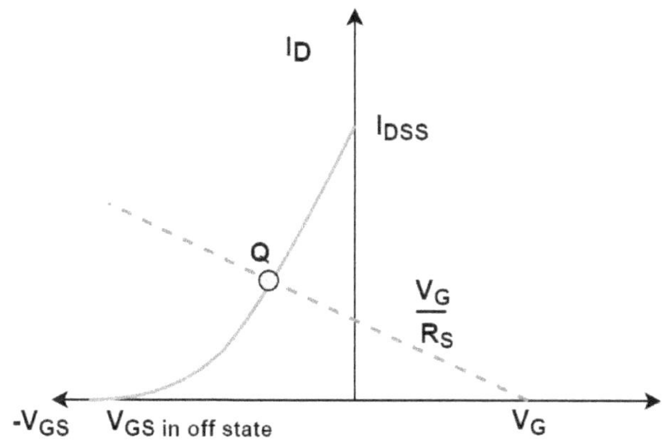

3.11 Current Source Bias

Current-source bias is a method for increasing the Q-point stability of a self-biased JFET by making the drain current essentially independent of V_{GS}. This is accomplished by using a constant-current source in series with the JFET source, as shown in figure.

In this circuit, a BJT acts as the constant-current source because its emitter current is essentially constant if $V_{EE} \gg V_{BE}$. A FET can also be used as a constant-current source. As you can see in figure, remains constant for any transfer characteristic curve, as indicated by the horizontal load line.

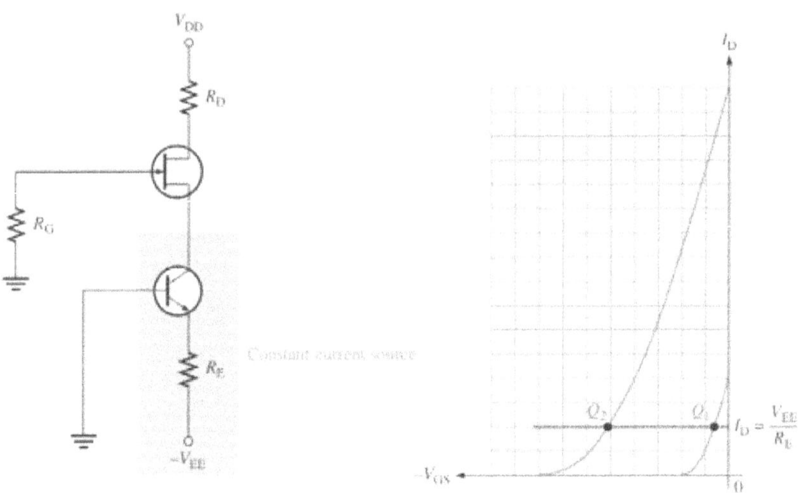

3.12 Mosfet Biasing

Feedback Biasing for Enhancement mode MOSFET

The circuit diagram for feedback biasing is shown. Drain is connected to gate through feedback bias resistor R_G. As there is no gate current, there is no voltage drop across R_G and $V_{DS} = V_{GS}$. So drain and gate potentials are same w.r.t. ground source. This means bias point can only change by varying R_L.

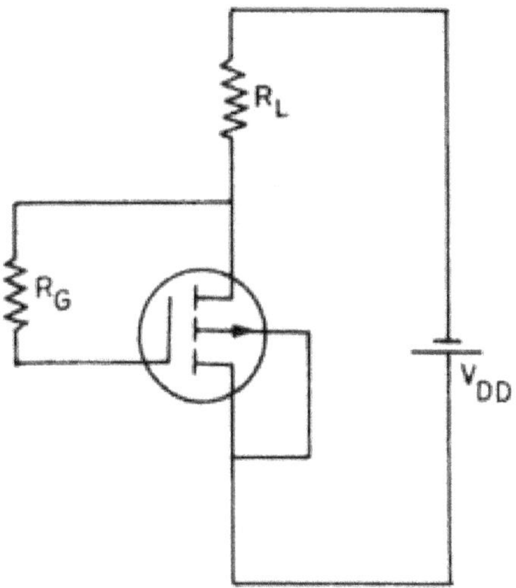

Fixed Biasing for enhancement MOSFET

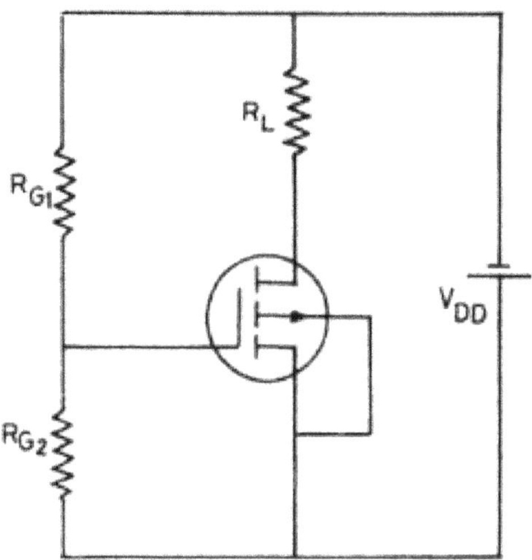

Two resistors R_{G_1} and R_{G_2} are used as voltage divider and gate-source voltage is fixed.

$$V_{GS} = \frac{R_{G_2} \cdot V_{DD}}{\left(R_{G_2} + R_{G_2}\right)}$$

Self Biasing for Depletion MOSFET

This is just like JFET self Bias. However here R_G is larger in case of MOSFET. Also in MOSFET, gate to source voltage can also be $+ve$ so to bias V_{GS} must be $-ve$.

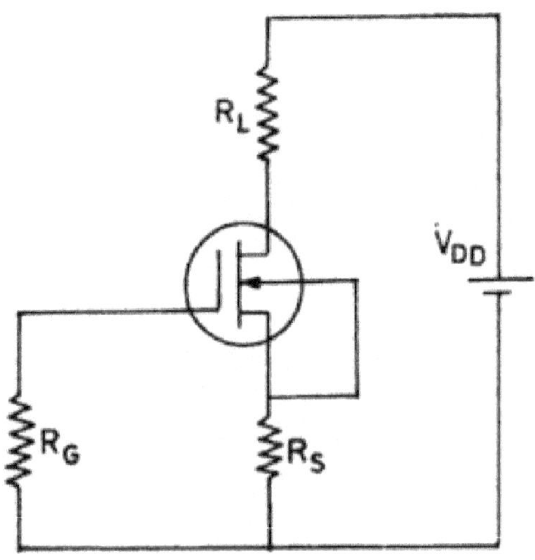

Fixed Biasing for Depletion MOSFET

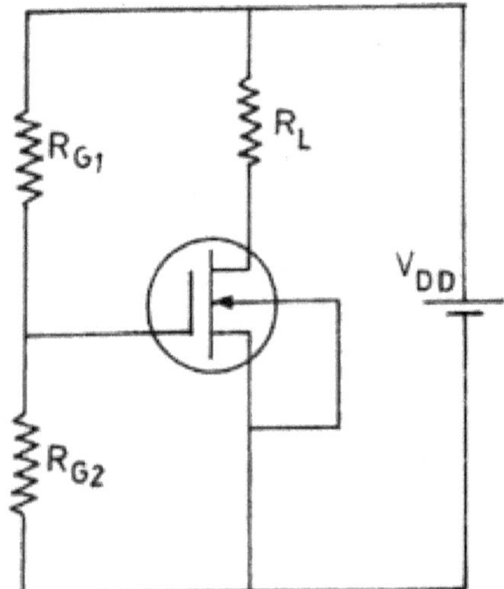

This is again as Enhancement MOSFET.

SUMMARY

- FETs are unipolar devices unlike BJTs

- At boundary between Ohmic and active region

$$V_{DS} = V_{GS} - V_P$$

- $I_D = I_{DSS}\left(\dfrac{V_{DS}}{V_P}\right)^2$

- **CS** configuration like the common emitter in BJT, the input is applied to the Gate terminal and its output is taken from the Drain

- **CG** configuration like the common base of BJT, the input is applied to the Source and its output is taken from the Drain with the Gate connected directly to ground

- **For small signal model of JFET**

- $g_m = transconductance = \dfrac{I_D}{V_{GS}}$

- $g_m = g_{m_0}\sqrt{\dfrac{I_D}{I_{DSS}}}$

- Depletion Type MOSFET – the transistor requires the Gate-Source voltage, (V_{GS}) to switch the gadget "OFF". The exhaustion mode MOSFET is equal to a "Regularly Closed" switch.

- Enhancement Type – the transistor requires a Gate-Source voltage, (V_{GS}) to switch the gadget "ON". The upgrade mode MOSFET is identical to an "Ordinarily Open" switch.

- If we tend to apply a little time-varying signal to the input, then under the proper circumstances the mosfet circuit will act as a linear electronic amplifier providing the transistors Q-point is somewhere close to the middle of the saturation region, and therefore the input is small enough for the output to stay linear.

- drain to source voltage in JFET self bias is

$$V_{DS} = V_D - V_S = V_{DD} - I_D(R_D + R_S)$$

- drain current in JFET voltage-divider bias is $I_D = \dfrac{V_G - V_{GS}}{R_S}$

SHORT ANSWER TYPE QUESTIONS

1. Find the overall transconductance of two transistors with input voltage is V_{bias} and V_i. If both MOSFET are saturated and for particular V_{out} g_m is $\dfrac{\partial I_{out}}{\partial V_i}$

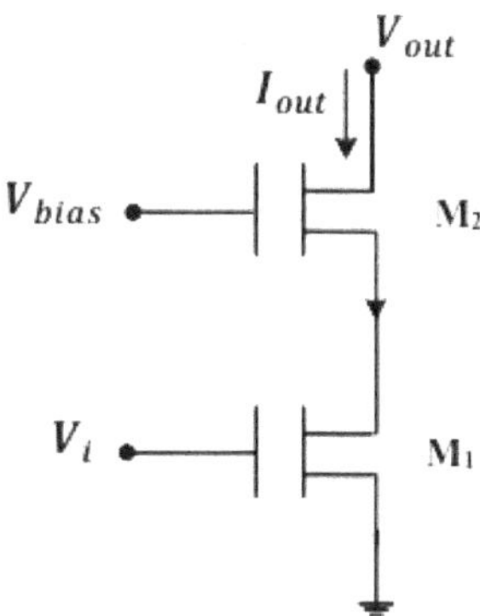

2. What would be the graph of transconductance to gate voltage at constant drain voltage if MOSFET is operated in linear region?

3. If an n-channel MOSFET has $V_{threshold} = 0.85V$ and V_{drain} is 2.1V. Now if this drain voltage is changed to 2V with initial drain current of 0.5 mA.

4. What would be the new drain current? The figure is shown of the circuit below

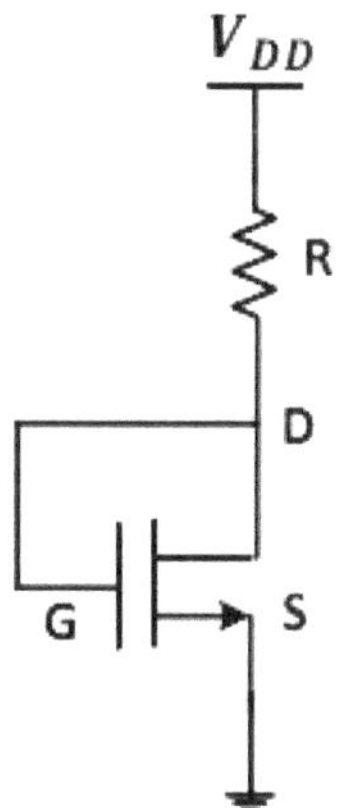

5. An n – channel JFET has pinch – off voltage
   $V_p = -5\,V, V_{DS}(max) = 20\,V,$ and $g_m = 2\,mA/V.$
   The min 'ON' resistance is achieved in the JFET for

6. The transit time of the current carries through the channel of a JFET
   decides its _____________ characteristic

7. In the C MOS inverter circuit shown if the transconductance parameters
   of N MOS and P MOS transistor are

$$K_n = K_p = \mu_n C_{OX} \frac{W_n}{L_n} = \mu_p C_{OX} \frac{W_p}{L_p} = 40\,\mu A/V^2$$

   and threshold voltages are $V_T = 1V$ the current I is

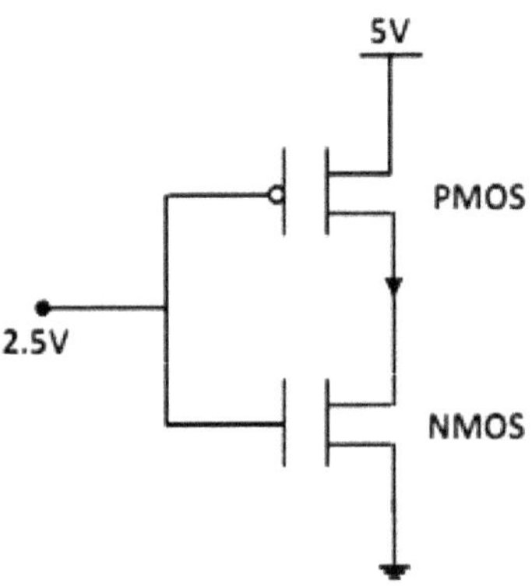

8. An n – channel JFET has $I_{DSSS} = 2\,mA$ and $V_p = -4\,V.$ It's
   transconductance $g_m (in\ mA/V)$. An applied GATE to source voltage
   V_{GS} of – 2V is

Chapter 4

Feedback Amplifiers

A Brief Chapter Overview

4.1	Feedback Amplifiers
4.2	Classification of Amplifiers
4.3	Gain of different types of amplifier-feedback configurations
4.4	Hartley Oscillator
4.5	Types of oscillators
4.6	RC Phase Shift Oscillators
4.7	Colpitts Oscillators

4.1 Feedback Amplifiers

When we say feedback, we mean the portion of output voltage is released back to the input voltage. This means that input voltage is changed successively. Depending on the feedback voltage, the input voltage can be increased or decreased if feedback voltage is added or subtracted from input voltage.

There are some benefits of feedback in a circuit. These effects are

1. Noise cancellation circuits

2. Overall gain can be changed

3. Bandwidth of signal is increased

4. More stable circuits

5. To make oscillators that can convert DC voltage to AC voltages

6. Frequency response is increased

7. Linear circuits

8. Higher input impedance and output impedance is low

Basic Feedback Circuit

Input signal V_s goes to amplifier A, where amplification process occurs and output signal V_0 is taken away from outer circuit. Output signal is now multiplied by factor β. This factor is called feedback factor. This feedback factor is added to or subtracted from input signal in positive or negative feedback circuits.

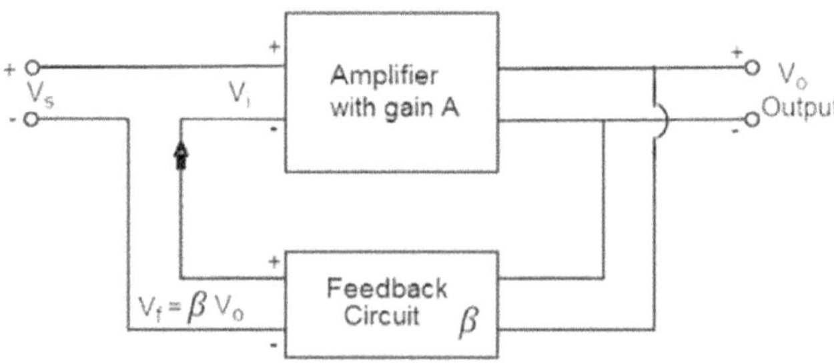

Fig. 4.1

Negative Feedback

In negative feedback, the feedback voltage is subtracted from input voltage. This means that feedback voltage is out of phase with input voltage and this also effect is decreased overall gain

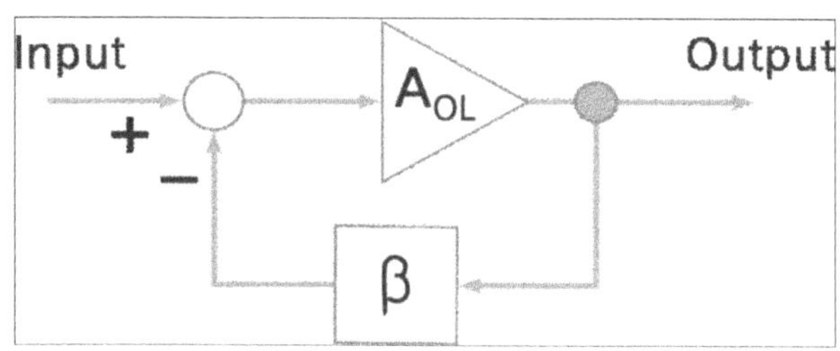

Fig. 4.1

This means that new input voltage after feedback is

Or $V_i = V_S - V_f$

$V_S = V_i + bV_o$

As gain of amplifier is

$$A = \frac{V_o}{V_i}$$

So in this way

$V_o = AV_i$

$V_S = (1 + \beta A)V_i$

New gain of amplifier circuit is A_f called the gain with feedback.

$$A_f = \frac{A}{1 + \beta A}$$

Also remember that without feedback $V_S = V_i$ so gain without feedback is $A = V_0/V_i$

Now,

$$A_f = \frac{A}{\beta A} = \frac{1}{\beta}$$

Notice that $A_f < A$. Also that V is a complex quantity and so voltage gains are complex too so $|A_f| < |A|$.

Also if $|\beta A|$ is way larger than 1, then

$$A_f = \frac{A}{\beta A} = \frac{A}{\beta}$$

Positive Feedback

In positive feedback, feedback voltage is in phase to the input voltage. In positive feedback condition the gain of the circuit is always greater than the gain without feedback

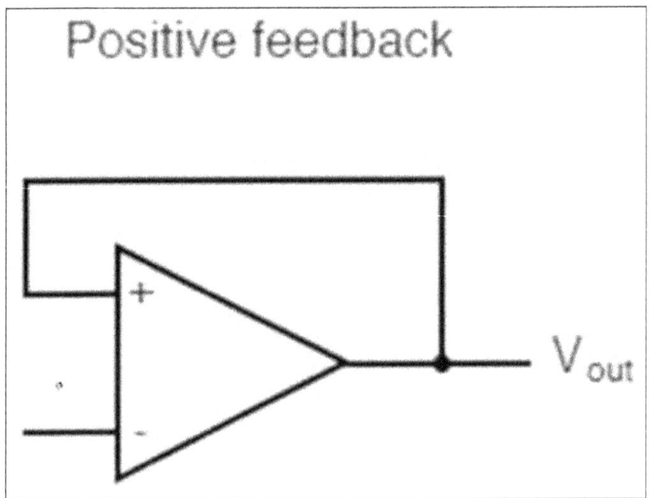

Fig. 4.3

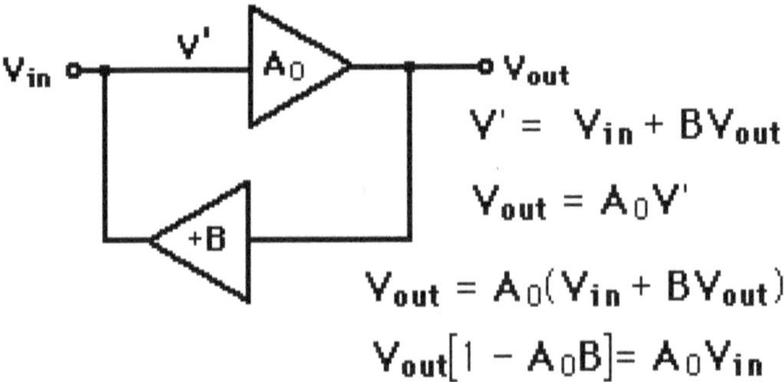

Here

$$V_i = V_S + V_f$$

Or

$$V_S = V_i - bV_o$$

$$V_S = (1 - \beta A)V_i$$

In case of positive feedback gain of the amplifier circuit is

$$A_f = \frac{A}{1 - \beta A}$$

This means overall gain of the amplifier is increased. Now look at the factor βA. If $|\beta A|$ is way larger than 1, then $A_f = -1/\beta$

Now if $|\beta A| = 1$, then $A_f = \infty$. It is only possible if even without any input voltage the output energy keeps flowing in the circuit. This is very important as at later stage we will exploit this property of amplifiers to make oscillators. Oscillators are circuits that can be used to make AC voltages out of DC voltages i.e., opposite to rectifiers.

4.2 Classification of Amplifiers

Based on the magnitudes of input and output impedances with respect to the source and load impedances an amplifier can be classified in four types:

1. Current Amplifier

In Norton equivalent circuit of amplifier one can see that if input resistance R_i of amplifier is very close to zero and far lesser than source resistance R_s. Then this type of amplifier is current amplifier. In this amplifier the source current I_s is equal to the input current I_i. In the output side of a current amplifier output resistance is R_0 is infinite and load resistance is very small than output resistance then output current is equal to the load current or $I_0 = I_L = A_i I_i$. This A_i

is called the current gain and does not depend on the source or load resistances

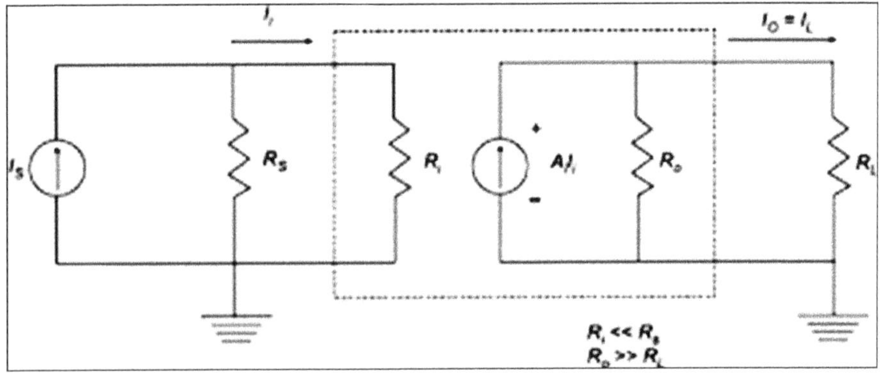

Fig. 4.4

2. Voltage Amplifier

In the venin equivalent circuit of the voltage amplifier one can see that input resistance R_i is infinite and source resistance R_s is very small thehn source voltage V_s is equal to the input voltage V_i. Also in the output side of voltage amplifier the load resistance is very large opposed to what be seen in the current amplifier. So that output voltage $V_o = A_v V_i$. The proportionally constant A_v is is called the voltage gain of amplifier and does not depend on source and load resistances. This type of amplifier is called the voltage amplifier.

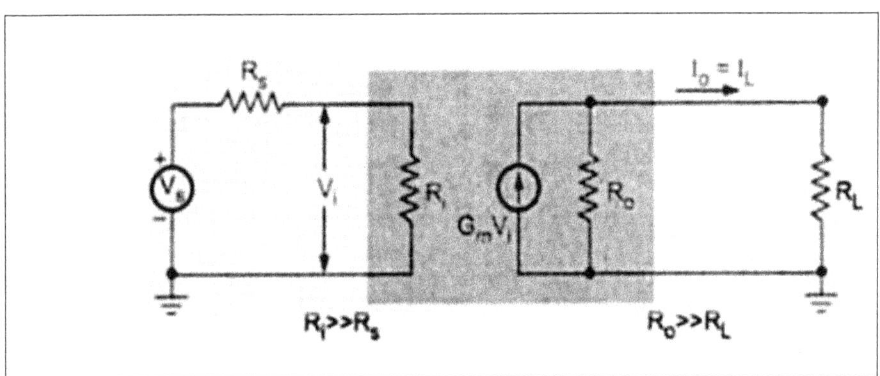

Fig. 4.5

3. Transconductance Amplifier

Transconductance amplifier is produced when the input in the venin circuit of voltage amplifiers joined with the output Norton circuit of current amplifier. So in this amplifier the output current is proportional to the input voltage and proportionality constant G_M is called transconductance gain and does not depend on input source and load resistances.

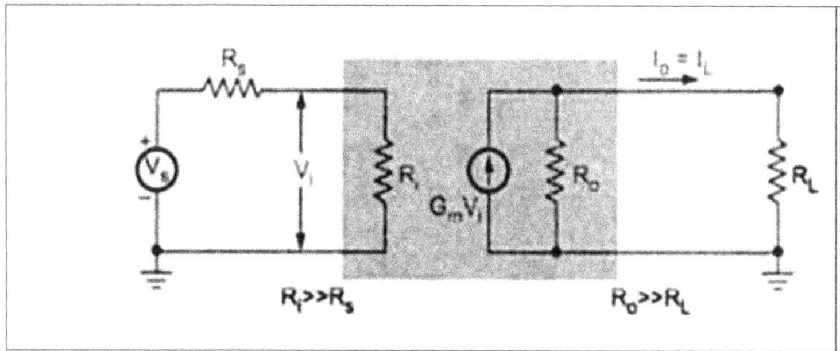

Fig. 4.6

4. Tran Resistance Amplifier

When Norton equivalent circuit of input of current amplifier is joined with the output the venin voltage amplifier circuit the this type of amplifier is called transresiatnce amplifier and then output voltage is written as $V_0 = R_M I_i$. Here R_m is called transresistance transfer ratio or simply transresistance gain of amplifier. This gain again is independent of source and load resistances.

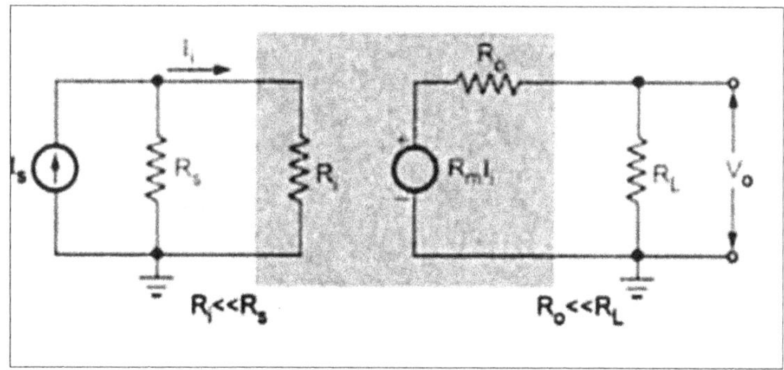

Fig. 4.7

Amplifier Circuit Components

To make an amplifier one needs the following components

    a. Signal source

    b. Mixer circuit

    c. Sampling network

    d. Feedback network

    e. Amplifier network

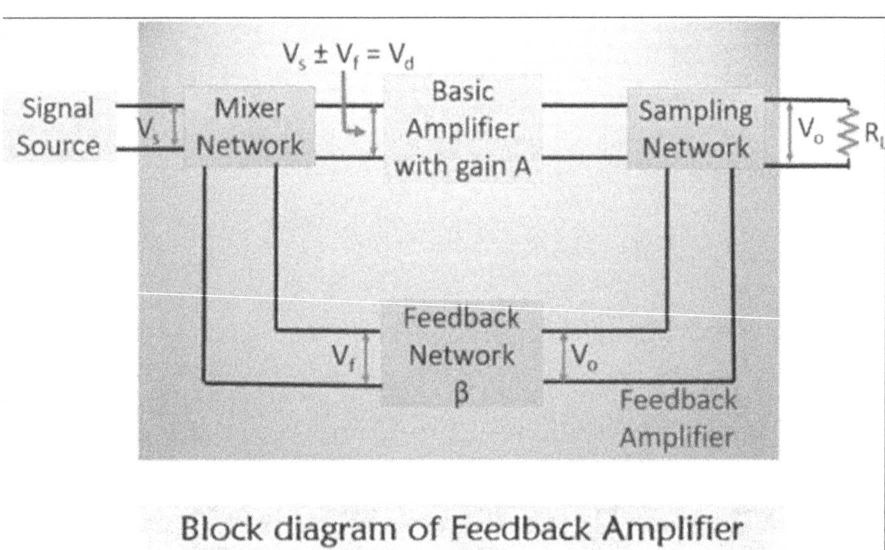

Block diagram of Feedback Amplifier

Here signal source would simply mean that any voltage or current source and amplifier network are all the amplifier types that we seen previously. These types are current, voltage, transresistance and transconductance amplifiers.

Mixer Network

When the output signal of amplifier is fed back to the input signal to amplifier there are two ways of mixing these signals one way is in series to the input source signal and other way is in shunt to the source.

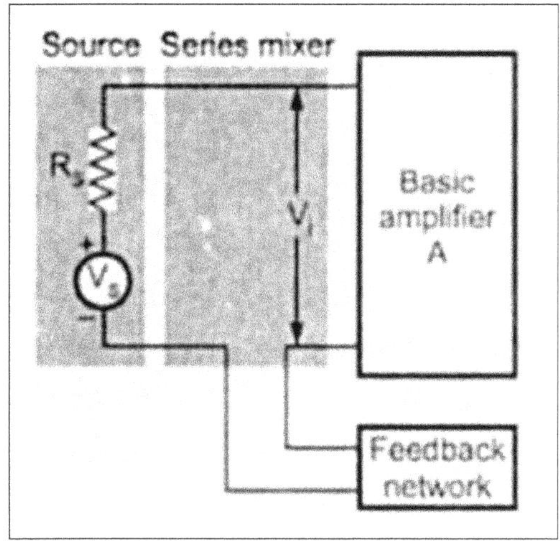

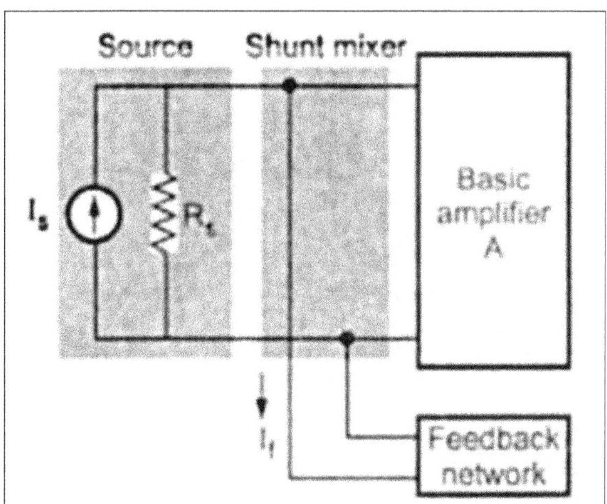

Sampling Network

The output signal from amplifier is taken from amplifier and given to the feedback circuit; there are two ways of taking these signals. One way is when feedback circuit is connected in series to the output of amplifier and other way that it is connected in shunt. When it is connected as shunt then it's called as voltage or node sampling to get maximum voltage to the feedback circuit and when it is connected as series it is called current or loops sampling to get maximum current to the feedback circuit.

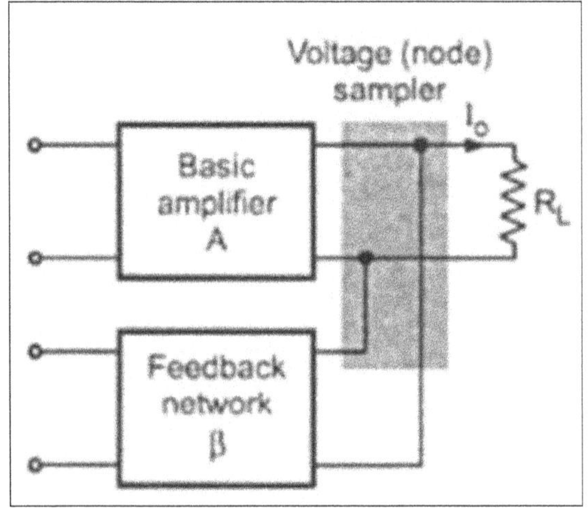

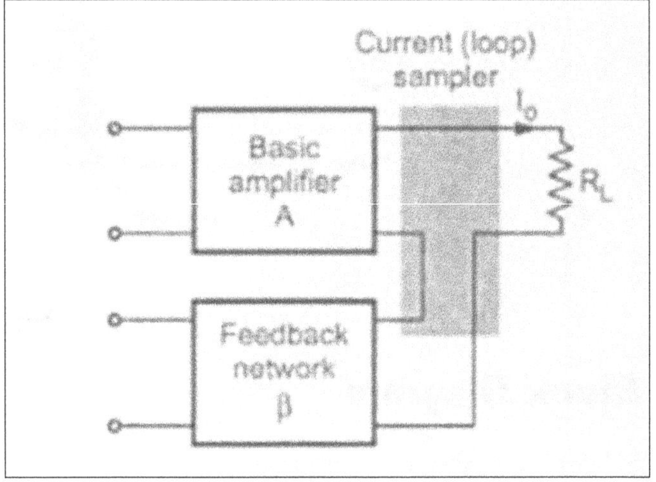

Fig. 4.9

Feedback Circuit

Any circuit that cuts off some portion of output signal and add this portion to the input signal is called feedback circuit. The feedback voltage that is mixed with input voltage i.e., $V_s = V_i + V_f$, V_f is called as feedback voltage.

$$V_f = \beta V_0$$

β here called as feedback factor of feedback circuit and it lies between 0 and 1.

Gain of an Amplifier with and without Feedback

We have seen some proportionally contants when we studies the classification of amplifiers.

There are voltage gain

$$A_v = \frac{V_o}{V_i}$$

Current gain

$$A_i = \frac{I_o}{I_i}$$

Transconductance gain

$$G_m = \frac{I_o}{V_i}$$

Transresistance gain

$$R_m = \frac{V_o}{I_i}$$

These are the gain of amplifier circuit when there is no feedback circuit. In the presence of a feedback circuit these gains are written as

Voltage gain with feedback

$$A_{if} = \frac{I_o}{I_S}$$

Current gain with feedback

$$A_{gf} = \frac{I_o}{I_S}$$

Transconductance gain with feedback

$$G_{mf} = \frac{I_o}{V_S}$$

Transresistance gain with feedback

$$R_{mf} = \frac{V_o}{I_S}$$

Remember from the classification of amplifier that in the absence of feedback circuit $V_S = V_i$ and $I_i = I_S$

Feedback Connection Types

There are four ways of connecting a feedback to the amplifier circuit in both input and output of amplifier. Series and parallel from output side and series and parallel from input side.

1. Voltage amplifier–series feedback

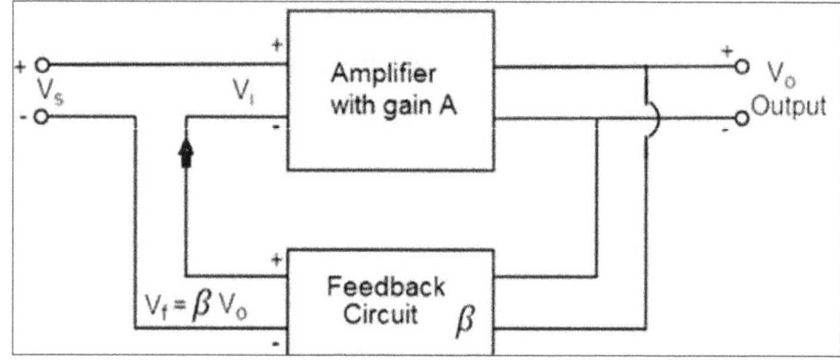

Fig. 4.10

2. Voltage amplifier – shunt feedback

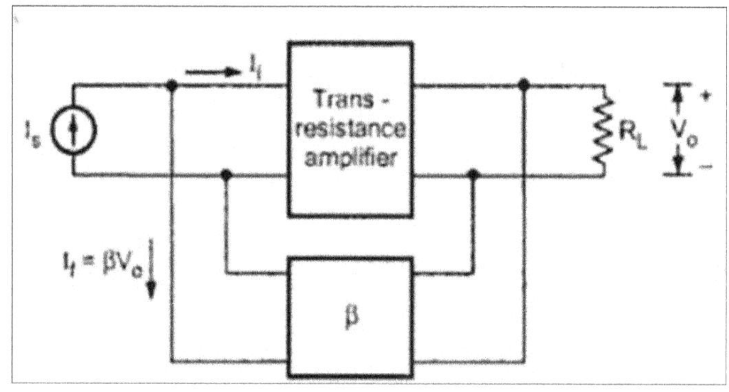

Fig. 4.11

3. Current amplifier-shunt feedback

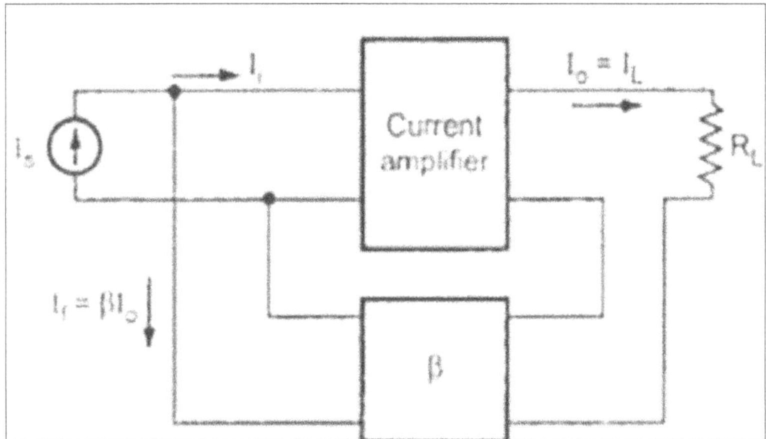

Fig. 4.12

4. Current amplifier-series feedback

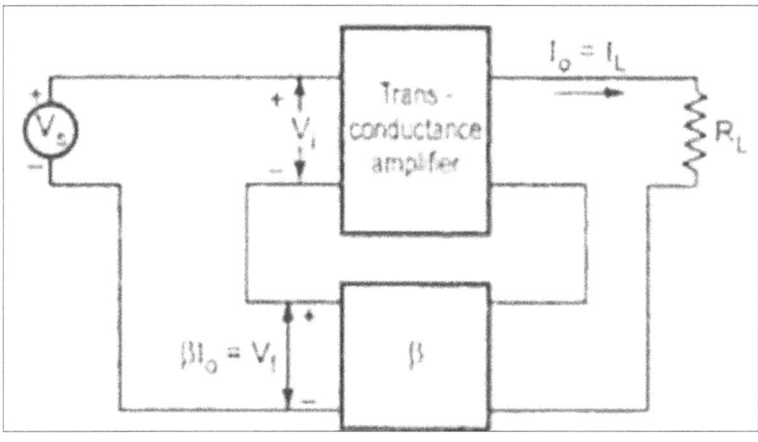

Fig. 4.13

These are represented diagrammatically; here one need to remember that voltage amplifier amplifies voltage so feedback is taken from output of amplifier is in the form of voltage. And voltage is same in parallel. So feedback is always connected in parallel to output of voltage amplifier. On the same basis a current amplifier always amplifies current so we require having currentfrom output of amplifier so feedback is connected in series to the output of amplifier.

Now when talking about input connection of feedback, a feedback can be connected in series or in parallel to the input of voltage and input of current amplifier. As both transresistance amplifier too amplifies voltage and transconductance amplifies current, so we can place them in voltage and current amplifiers. If feedback is connected so that voltage amplifier amplifies voltage from input current then it's a transresistance amplifier and if current amplifier amplifies current from input voltage then its transconductance amplifier.

This means if feedback circuit is passing voltage to the voltage amplifier input then it's a voltage amplifier. This feedback must be used in series to the input of amplifier as voltage always adds or subtracts in series and remain same in parallel. So this type of amplifier is also called as voltage amplifier-series feedback. On the other hand if feedback is passing current to the input of voltage amplifier then it's a transresistance amplifier. This feedback must be used in parallel to the input circuit of amplifier, as current always adds in parallel and remain same in series. So this amplifier is also called voltage amplifier-shunt feedback. Same kind of analogy can be used for current amplifier.

4.3 Gain of Different Types of Amplifier-Feedback Configurations

1. Gain of voltage amplifier-series feedback

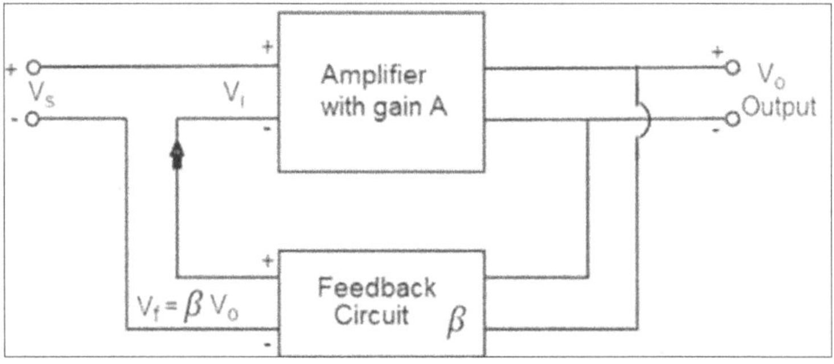

Fig. 4.14

$$A_{vf} = \frac{V_o}{V_S}$$

$$V_S = V_i + V_f$$

Where Vi and Vf are amplifier input and feedback voltage. Vi is the voltage of source connected to the amplifier in the absence of feedback circuit. It appears that new source of amplifier after connection of amplifier is Vs

Also that

$$V_f = AV_i$$

Where A is the gain of amplifier without feedback

$$A_{vf} = \frac{V_o}{V_i(1 + \beta A)}$$

And input impedence can be represented as

$$I_i = \frac{V_i}{Z_i} \; (V_S - V_f) \; \frac{1}{Z_i}$$

Or

$$I_i = V_S - \beta AV_i \times \frac{1}{Z_i}$$

$$I_i Z_i = V_i - \beta \, AV_i$$

$$V_S = I_i Z_i + \beta \, AV_i + bAZ_i I_i$$

Now,

$$Z_{if} = \frac{V_S}{I_i} = Z_i \; (1 + \beta A)$$

2. Gain of voltage amplifier-shunt feedback

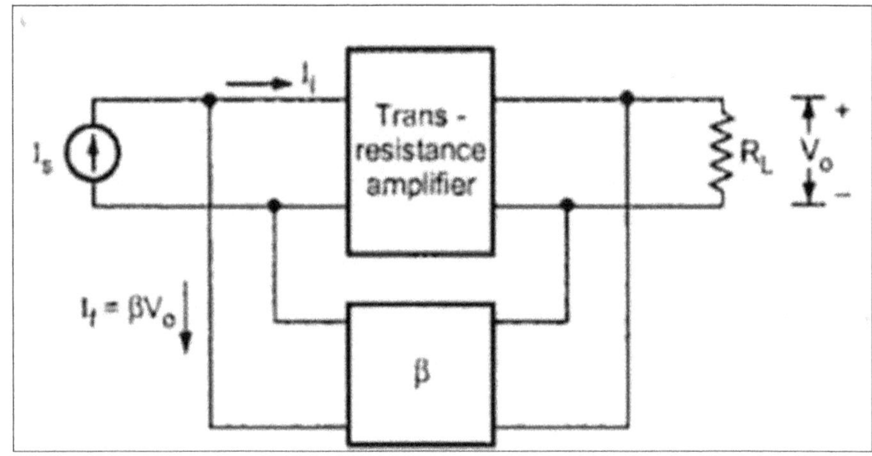

Fig. 4.15

We have seen that this amplifier is also called transresistance amplifier.

$$R_{mf} = \frac{V_o}{I_S}$$

$$I_S = I_i + I_f$$

Where, I_f is bV_o

And

$$V_o = R_m I_i$$

Where R_m is gain without feedback

So

$$R_{mf} = \frac{R_m}{1 + \beta R_m}$$

And input impedance can be represented as

$$Z_{if} = \frac{V_i}{I_S} = \frac{V_i}{I_i + I_f}$$

This means that

$$Z_{if} = \frac{V_i}{I_i + \beta V_o} = \frac{Z_i}{I_i + \beta R_m}$$

Output impedance of different amplifier-feedback Configuration

1. Voltage-series feedback

In this configuration output resistance can be measured by shorting theinput source or $V_s = 0$ and open the load resistance

If in absence of source output voltage of amplifier is V Then

$$V = IZ_o + AV_i$$

As

$$V_S = V_i + V_f$$

and if $V_S = 0$ then

$$V_i = - V_{if}$$

So

$$V_i = IZ_o - AV\beta$$

Or

$$V(1 + \beta A) = Iz_o$$

and

$$Z_{iof} = \frac{V}{I} = \frac{Z_o}{1 + \beta A}$$

2. Current amplifier-series feedback

Apply signal V to the output with shorting V_s. The output current will be I As $V_s = 0$

Then $V_i = V_f$

$$I = \frac{V}{Z_o} - G_m V_i = \frac{V}{Z_o} - G_m V_f$$

Or

$$I = V/Z_o - G_m \beta I$$

$$V = Z_o I (1 + G_m \beta)$$

And

$$Z_{of} = \frac{V}{I} = Z_o (I + \beta G_m)$$

Oscillators

Oscillator is any circuit that oscillates the signal with desired fixed amplitude and frequency. In other words Oscillators converts DC signal to AC signal opposite to rectifiers.

Oscillators don't require the input signal and employs the positive feedback circuit to create the oscillating circuit. Oscillators are very important tool for electronic circuits that are responsible for back conversion of signals from battery source. They are used in inverters

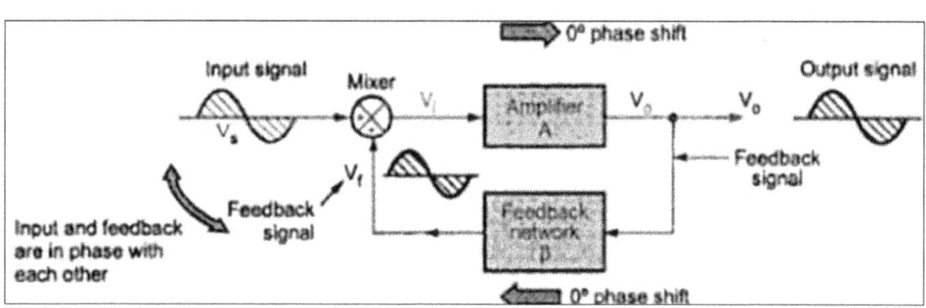

Fig. 4.16

As we said that the oscillators employs the positive feedback amplifier tocreate the oscillating output. We can see how

As gain of the amplifier circuit is given by

$$A = \frac{V_o}{V_i}$$

This is the gain of amplifier when no feedback is applied. This is also called theopen loop gain of the amplifier

When the circuit is connected in the feedback with feedback voltage V_f and applied voltage is V_s. In this way the total input voltage to the circuit is give by:

$$V_i = V_s + V_f$$

Where we told earlier in chapter that feedback voltage V_f is five by

$$V_i = \beta V_o$$

Again where V_o is the output voltage.

$$V_i = V_s + \beta V_o$$
$$V_S = V_i + \beta V_o$$

So the gain of amplifier would become as given by

$$A_{vf} = \frac{V_o}{V_i - \beta V_o}$$

This can also be written as

$$A_{vf} = \frac{A}{1 - A\beta}$$

If $A\beta$ is 1. This results in infinite gain of the circuit, means that with zero input voltage there will be some output voltage. This forms the basis of oscillators. In real circuits β need not be real. They are imaginary. So $|A\beta| = 1$ is often called the barkhausen criteria of oscillation.

Barkhusen Criteria

It states that

1. With zero input voltage the feedback voltage acts as the input voltage.

2. The total phase shift when voltage enters the loop to the time when voltage exits the loop undergoes a $2 * \pi$ rotation. This means voltageencounters a phase shift of 0 to 2π.

3. The magnitude of product of open loop gain and the feedback gain isunity.

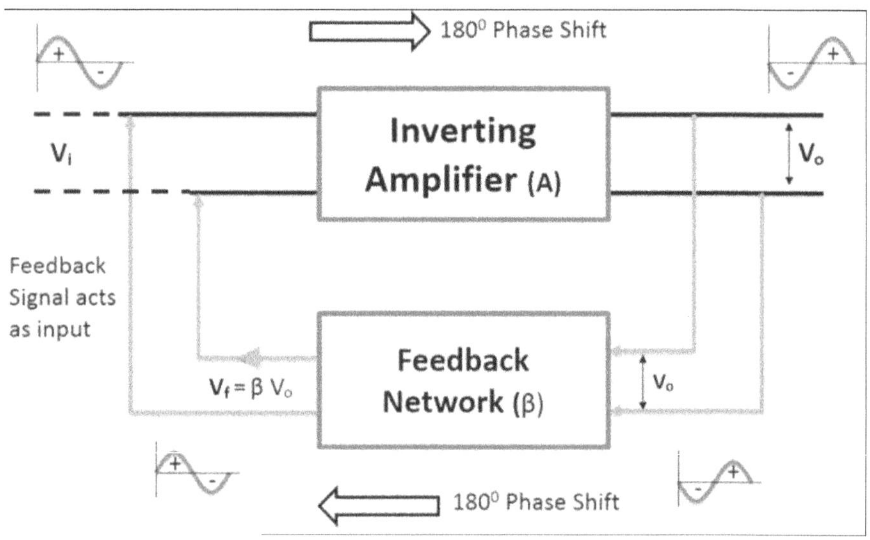

Fig. 4.17

This barkhausen criteria is the very imaginary kind of circuit. In real world the there needs to be a small current no matter how small, must not be equal to zero. If the product $|A\beta|$ greater than 1 then this results in growing of oscillations.

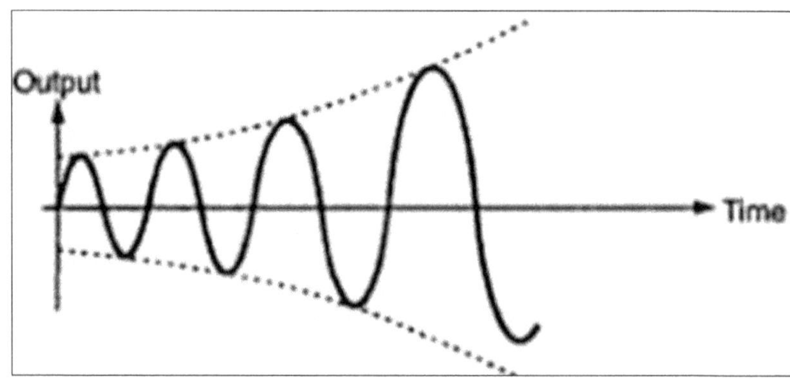

However the barkhausen criteria leads to sustained oscillations. In real world circuits we need to provide a small value of input voltage at regular intervals. So the voltage does not go damped.

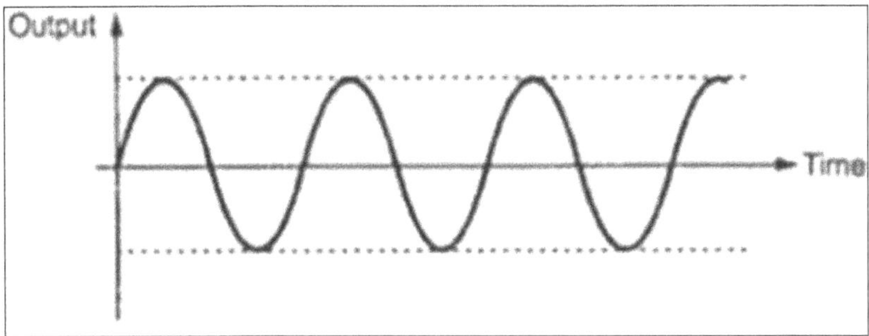

When $|A\beta|$ is less than 1. This will result in damped oscillations with time. Aftera long time the signal will eventually decay.

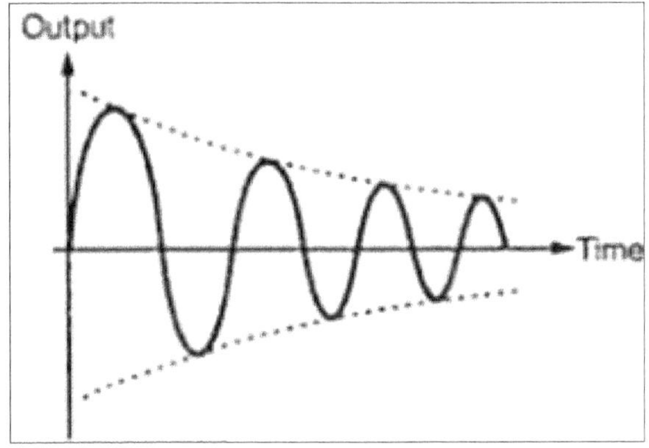

4.4 Types of Oscillators

1. Based on output waveform

They are called sinusoidal and non-sinusoidal oscillators. Non-sinusoidal aresquare wave, triangular, saw-tooth shaped and many more.

2. Based on The circut components

Some oscillators contain on resistors and capacitors. So we have RC oscillators,others have L and C so we call them LC oscillators.

3. Based on Operating frequency

They are called low frequency and High frequency oscillators. LF and HF. Some are referred as Audio Frequency AF and some others ss Radio frequency RF

Few oscillator circuits:

4.5 RC Phase Shift Oscillators

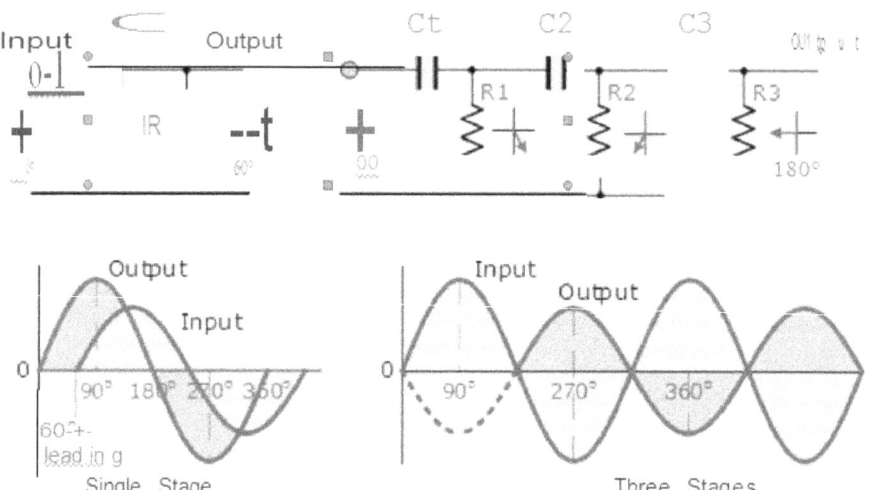

The circuit on the left shows a single resistor -capacitor network whose output voltage "leads the input voltage by some angle less than 90°. An ideaII single-pole RC circuit would produce a phase shift of exactly 90°, and because 180 P of phase shift is required for oscillation, at least two single -poles must be used in an RC *oscillator* design.

However in really it is d if f11cult to obtain exactlly 90 of phase shift so more stages are used. The amount of actual phase shift in the circuit depends upon the values of the resistor and the capacitor, and the chosen frequency of oscillations with the phase angle (¢>) being gi1v en as:

RC Phase Angle

$$\frac{1}{2rr/C} \quad R \quad R$$

$$Z = \sqrt{R^2 + (XC)^2}$$

$$\tan^{-1} \frac{Xe}{R}$$

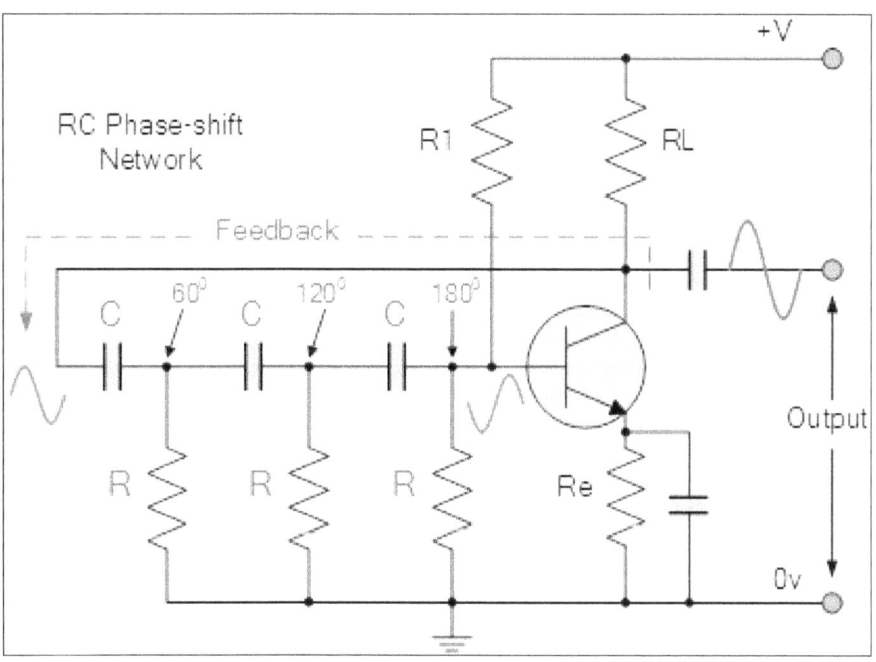

The basic RC Oscillator which is also known as a Phase-shift Oscillator, produces a sine wave output signal using regenerative feedback obtained from the resistor-capacitor combination. This regenerative feedback from the RC network is due to the ability of the capacitor to store an electric charge, (similar to the L tank circuit). This resistor-capacitor feedback network can be connected as shown above to produce a reading phase shift (phase advance network) or interchanged to produce a lagging phase shift 'phase retard network) the outcome is still the same as the sine wave oscillations only occur at the frequency at which the overall phase-shift is 360° By varying one or more of the resistors or capacitors in the phase-shift network, the frequency can be varied and generally this is done by keeping the resistors the same and using a 3-ganged variable capacitor. fall the resistors, Rand the capacitors, C in the phase shift network are equal in value, then the frequency of oscillations produced by the RC oscillator is given as:

$$f_r = \frac{1}{2\pi RC \sqrt{2N}}$$

Here R stands for resistance for capacitor and N for number of RC stages, N=3

Wein Bridge Oscillator

Wien Bridge oscillator is two stages RC coupled amplifiers. Generally an oscillator has a phase shift of input voltage of 180 from the amplifiers and other 180 from the feedback circuit. This is the barkhausen criteria of oscillations. However in Wien bridge oscillator non-inverting amplifiers are used to introduce zero phase shifts from the amplifier stages. This means that this type of oscillators has zero phase shifts.

The wien bridge oscillator has the two RC coupled amplifiers that are placed inparallel. This is shown in figure

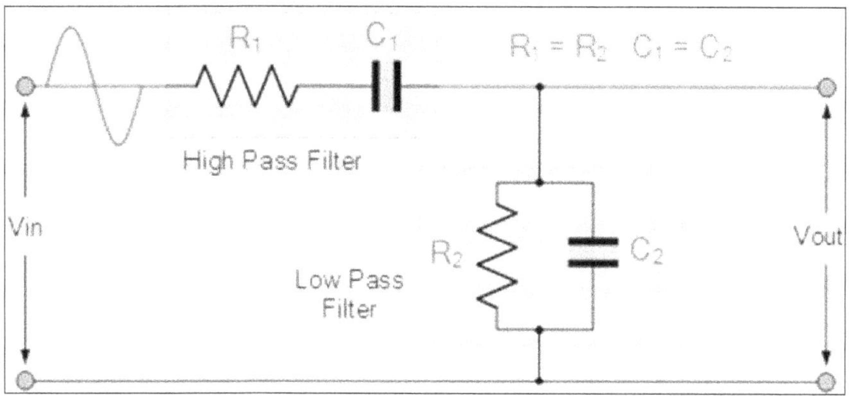

So the resistors and capacitors R_1 and C_1 forms a high pass filter and R_2 and C_2 forms a low pass filter. So wien bridge oscillators a specific bandwidth across a frequency of resonance f_r.

This way a wien bridge oscillator provides more stability by working in its specific frequency bandwidth and has less noise in its input.

At low range of frequencies high pass filter blocks any input signal and during low frequencies low pass filter blocks the signal. This is because the reactance of capacitor 1 becomes too high and reactance of Capacitor 2 becomes too low for high and low frequencies.

This type of network is called lead-lag network as at high frequencies it acts as lag while at low frequencies it leads.

Oscillator Output Gain and Phase Shift

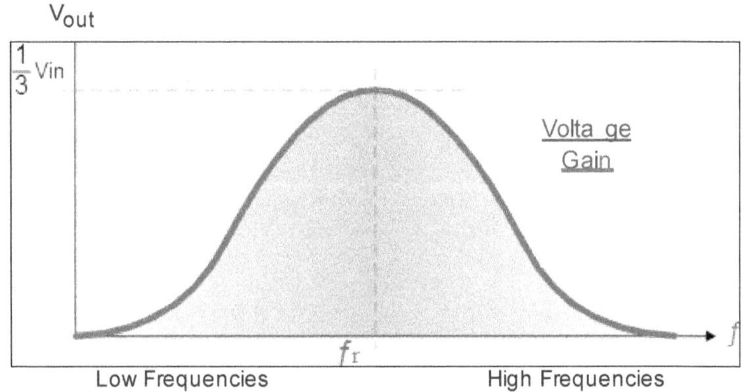

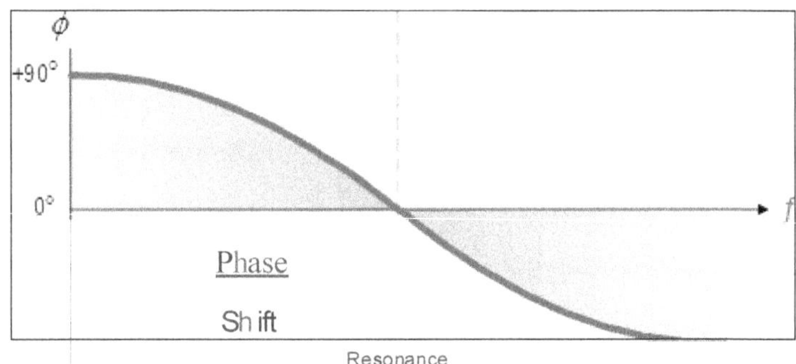

Wien Bridge Oscillator Frequency

$$fr = \frac{1}{21\ tRC}$$

Where,

fr is the Resonant Frequency in Hertz

R is the Resistance in Ohms.

C is the Capacitance in Farads

In wien bridge oscillator $\beta = \dfrac{1}{3}$. So A must be greater than of equal to 3. So in this case $V_0 \geq 3\ V_{in}$.

4.5 Hartley Oscillator

It is an LC oscillator that has two inductors in series with a parallel capacitor. The circuit elements L and C are called tank elements or oscillatoryelements. These oscillators have high frequency ranges.

Operation of LC oscillator

The basic operation of LC oscillator is that when at time = 0, the capacitor getscharged by the electrostatic energy.

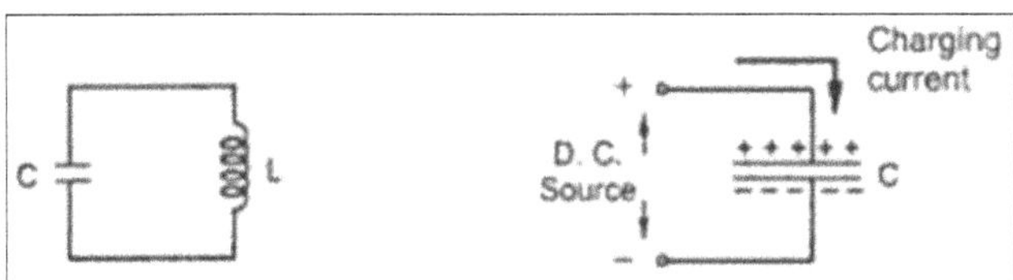

This capacitor is now connected in parallel to the inductor L. And capacitor eventually drains its energy to the inductor L. As a result the magnetic field surrounds the inductor. And as soon as the capacitor fully drains of its stored potent. The inductor gets fully stored of em energy.

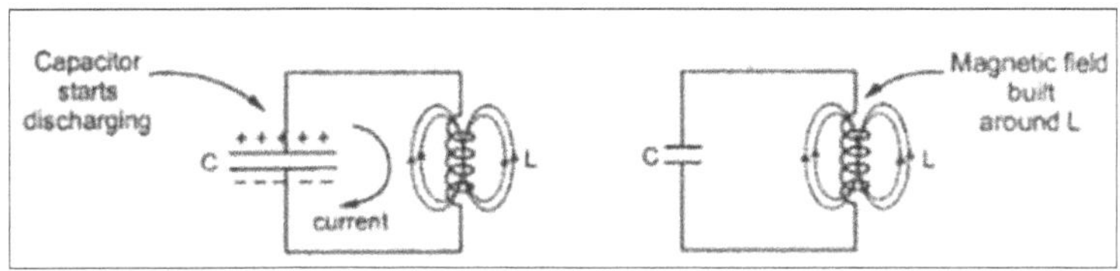

Now the inductor begins to lose its energy in the form of potential that can begin charging the capacitor now. The capacitor is charged in opposite directionas per lenz's law.

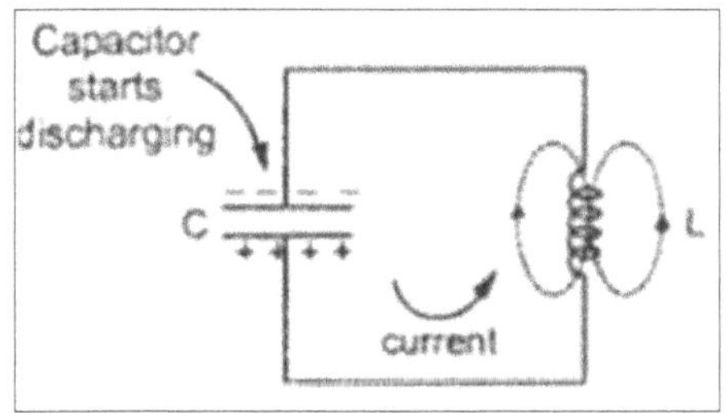

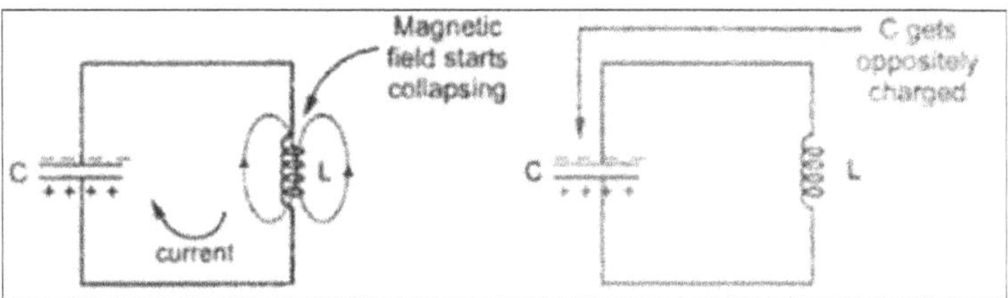

This process continues and inductor after draining its energy ready itself to get charged by the capacitor. This is again in opposite direction. So oscillator's works and polarities of current changes after each cycle of oscillation.

However this is a form of decaying oscillation. Every time the cycle completes and the other cycle begins. The next cycle has less energy than itsprevious cycle.

Operation of Hartley Oscillator

We have seen now that the LC oscillator has no means of controlling their amplitude of oscillation. Also if feedback is too high then the amplitude of oscillation will always goes on increasing. So it becomes difficult to tune LC circuits.

However it is possible to feedback exactly the right amount of voltage for constant Amplitude oscillation. If we feedback more than is necessary the amplitude of the oscillations can be controlled by biasing the amplifier in such a way that if the oscillations increase in amplitude, the bias is increased and the gain of the amplifier is reduced. If the amplitude of the oscillations decreases the bias decreases and the gain of the amplifier increases, thus increasing the feedback. In this way the amplitude of the oscillations are kept constant using a process known as Automatic Base Bias. One big advantage of automatic base bias in a voltage controlled oscillator, is that the Oscillator can be made more efficiency by providing a Class- B bias or even a class -c bias condition of the transistor. This has the advantage that the collector current only flows during part of the oscillation cycle so the quiescent collector current is very small. Then this self-tuning base oscillator circuit forms one of the most common types of LC parallel resonant feedback oscillator configurations called the Hartley Oscillator circuit.

Design of Hartley Oscillator Circuit

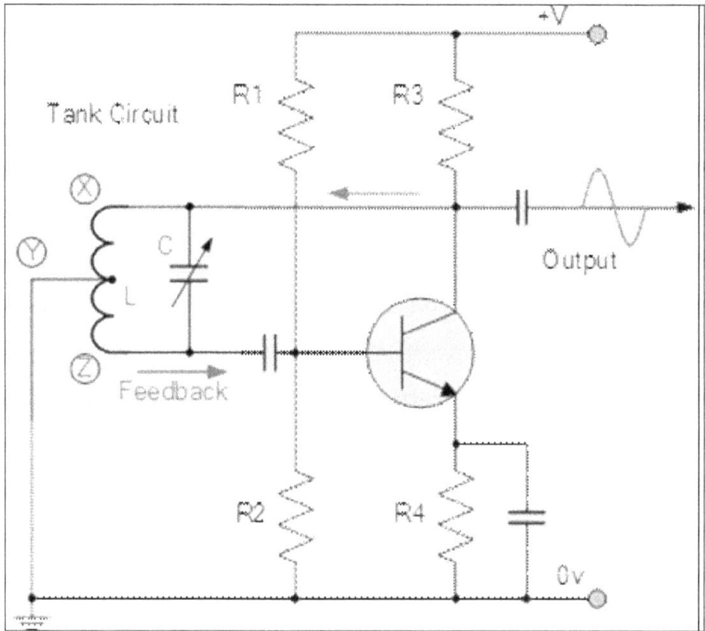

Fig. 4.19

When the circuit is oscillating the voltage at point X (collector), relative to point'emitter), is 180 out-of-phase with the voltage at point Z (base) relative to point Y. At the frequency of oscillation, the impedance of the Collector load is resistive and an increase in Base voltage causes a decrease in the Collector voltage.

Then there is a 180° phase change in the voltage between the Base and Collector and this along with the original 180% phase shift in the feedback loop provides the correct phase relationship of positive feedback for oscillations to be maintained.

The amount of feedback depends upon the position of the "tapping point" of the inductor.

If this is moved nearer to the collector the amount of feedback is increased, but the output taken between the Collector and earth is reduced and vice versa. Resistors, R1 and R2 provide the usual stabilizing DC bias for the transistor in the normal manner while the capacitors act as DC-blocking capacitors.

Hartley oscillators has frequency of

$$f = \frac{1}{2pi\sqrt{L_{eq}\,C}}$$

Where, L_{eq} is equivalent inductance

$$L_{eq} = L_1 + L_2 + 2M$$

Where M is the mutual inductance. If L1 and L2 are working collectively Mis positive otherwise M is negative.

4.7 Colpitts Oscillators

In colpitts oscillators the two capacitors are in series and the inductor isin parallel to the resistors.

Basic Design

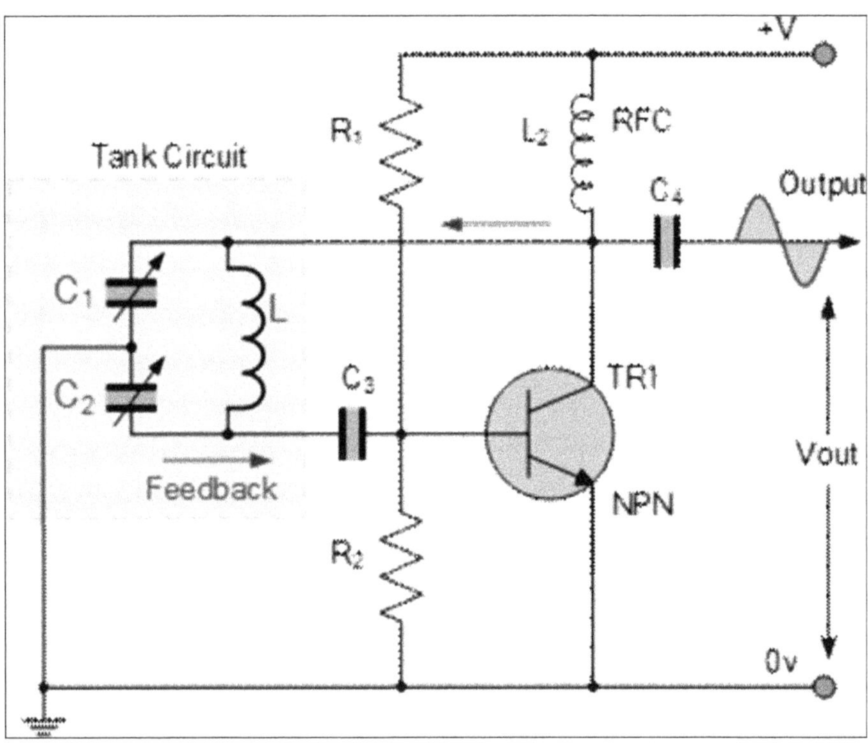

The emitter terminal of the transistor is effectively connected to the junction of the two capacitors, C1 and C2 which are connected in series and act as a simple voltage divider. When the power supply is firstly applied, capacitors C1 and C2 charge up and then discharge through the coil L. The oscillations cross the capacitors are applied to the base-emitter junction and appear in the amplified at the collector output. Resistors, R1 and R2 provide the usual stabilizing DC bias for the transistor in the normal manner while the additional capacitors act as a DC-blocking bypass capacitors. A radio-frequency choke (RFC) is used in the collector circuit to provide a high reactance (ideally open circuit) at the frequency of oscillation, (fr) and a low resistance at DC to help start the oscillations.

The required external phase shift is obtained in a similar manner to that in the Hartley oscillator circuit with the required positive feedback obtained for sustained undamped oscillations. The amount of feedback is determined by the ratio of C1 and C2. These two capacitances are generally ganged together to provide a constant amount of feedback so that as one is adjusted the other automatically follows.

The frequency of oscillations for a Colpitts oscillator is determined by the resonant frequency of the LC tank circuit and is given as:

$$f = \frac{1}{2\pi\sqrt{LC_T}}$$

Where C_T is the capacitance of C1 and C2 connected in series and is given as:

$$\frac{1}{C_T} \quad \frac{1}{C_1} + \frac{1}{C_2} \quad \text{or} \quad C_2 = \frac{C_1 \times C_2}{C_1 + C_2}$$

SUMMARY

- When we say feedback, we mean the portion of output voltage isreleased back to the input voltage

- In negative feedback, the feedback voltage is subtracted from inputvoltage.

- In positive feedback, feedback voltage is in phase to the input voltage

- Based on the magnitudes of input and output impedances with respect to the source and load impedances an amplifier can be classified in four types

- To make an amplifier one needs signal source, mixer circuit, sampling rule, feedback and Amplifier

- There are four ways of connecting a feedback to the amplifier circuit in both input and output of amplifier. Series and parallel from output side and series and parallel from input side

- Oscillator is any circuit that oscillates the signal with desired fixed amplitude and frequency

- Oscillators don't require the input signal and employs the positive feedback circuit to create the oscillating circuit

- Barkhusen criteria states that the total phase shift when voltage enters the loop to the time when voltage exits the loop undergoes a $2 * \pi$ rotation. This means voltage encounters a phase shift of 0 to2π

- Wien Bridge oscillator is two stage RC coupled amplifiers. Generally an oscillator has a phase shift of input voltage of 180 from the amplifiers and other 180 from the feedback circuit

- Hartley is a LC oscillator that has two inductors in series with a parallel capacitor. The circuit elements L and C are called tank elements or oscillatory elements. These oscillators have high frequency ranges

- In colpitts oscillators the two capacitors are in series and the inductor is in parallel to the resistors

SHORT ANSWER TYPE QUESTION

1. An amplifier has a gain of 2 × 105 without feedback. Determine the gain if negative voltage feedback is applied. Take feedback fraction mv = 0.02.

2. An amplifier has a gain of 10,000 without feedback. With negative voltage feedback, the gain is reduced to 50. Find the feedback fraction.

3. A feedback amplifier has an internal gain Av = 40db and feedback fraction mv = 0.05. If the input impedance of this circuit is 12 kΩ, what would have been the input impedance if feedback were not present.

4. Calculate the gain of a negative voltage feedback amplifier with an internal gain Av = 75 and feedback fraction mv = 1/15. What willbe the gain if Av doubles?

5. An amplifier with negative feedback has a voltage gain of 100. It isfound that without feedback, an input signal of 50 mV is required to produce a given output, whereas with feedback, the input signal must be 0.6 V for the same output. Calculate (i) gain without feedback (ii) feedback fraction.

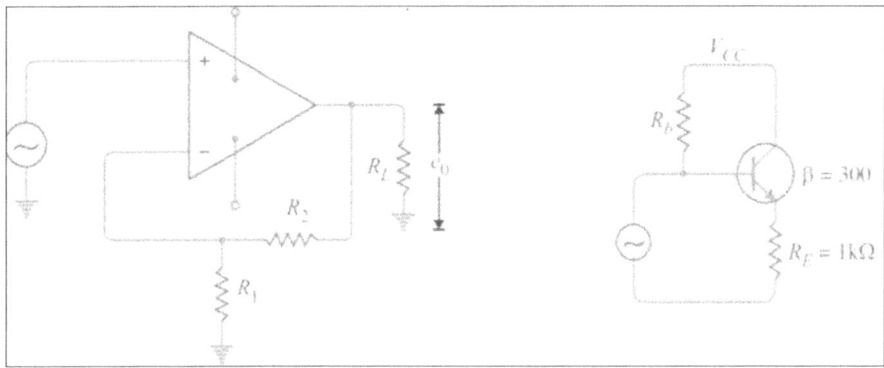

6. Fig. 13.30 shows the negative feedback amplifier. If the gain of the amplifier without feedback is 105 and R1 = 100 Ω, R2 = 100 kΩ, find (i) feedback fraction (ii) gain with feedback. [(i) 0.001(ii) 1000]Amplifiers with Negative Feedback.

7. In Fig. 13.31, if input and output impedances without feedback are 2 MΩ and 500 Ω respectively, find their values after negative voltage feedback.

8. An amplifier has a current gain of 240 without feedback. When negative current feedback is applied, determine the effective currentgain of the amplifier. Given that current attenuation mi = 0.015.

9. An amplifier has an open-loop gain and input impedance of 200 and 15 kΩ respectively. If negative current feedback is applied, what is the effective input impedance of the amplifier? Given that currentattenuation mi = 0.012. [4.41 kΩ] 10. An amplifier has Ai = 200 and mi = 0.012. The open-loop output impedance of the amplifier is 2kΩ.If negative current feedback is applied, what is the effective output impedance of the amplifier?

Chapter 5

Operational Amplifiers

A Brief Chapter Overview

5.1	Introduction	
5.2	Types of Op-Amp:	
5.3	Open Loop Operational Amplifiers	
5.4	Closed Loop Operational Amplifiers	
5.5	Parameters of Operational Amplifiers	
5.6	Ideal Op-Amp as integrator	
5.7	Ideal Op-Amp as differentiator	
5.8	Applications of Operational amplifiers	

OPERATIONAL AMPLIFIERS

5.1 Introduction

An Operational Amplifier commonly called Op-Amp is an integrated circuit of number of Bipolar Junction Transistors (BJT). It is a linear amplifying circuit commonly used in computers for mathematical operations like addition, subtraction and even integration and differentiation. By linear amplifying circuit we mean that it has linear relation between input voltage (V_i) and output voltage (V_o). Mostly Op-Amp is used in negative feedback circuit where its parameters like gain, impedance and bandwidth are dependent on external circuit elements. In negative feedback Op-Amp act as Amplifiers and in positive feedback it act as a Function Generator.

Symbol of Operational Amplifier

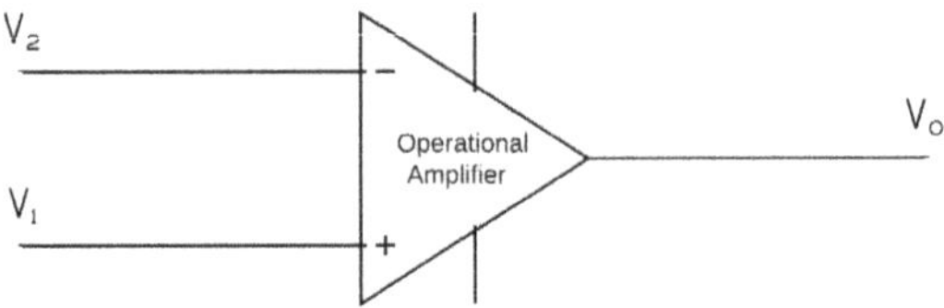

Fig. 1: Symbol of Op-Amp

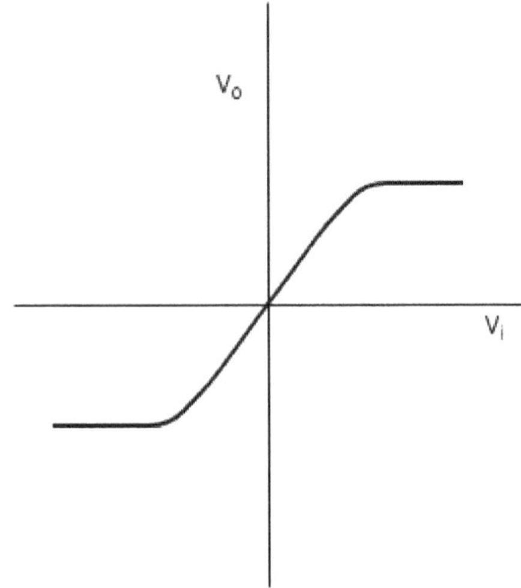

5.2 Types of Op-Amp

a. On Basis of Nature:

Ideal Op-Amp: Ideal Op-Amp has infinite gain that means $\dfrac{V_0}{V_i}$ is infinite. Even if no input voltage is supplied there is output. Some characteristics of ideal op-amp are:

1. Infinite gain

2. Infinite Input Impedance

3. Zero input current

4. Zero output impedance

5. Infinite Bandwidth

6. No phase change with frequency

7. No noise

8. Infinite Common mode rejection ratio

9. Zero input offset voltage and current

10. Infinite Power supply rejection ratio

Real Op-Amp: For real Op-Amp there is high voltage gain however not infinite and there is some value of input current. Output impedance too has a high value.

On Basis of Feedback:

Open Loop Op-Amp: Open loop Operational Amplifiers are the one that are used as comparators. They are simplest as output has no impact on its input. This means Open loop Op-Amp are not connected in a feedback circuit.

Gain of such Op-Amp is called Open loop gain and is denoted as A_{OL}

$$A_{OL} = \frac{V_0}{V_i}$$

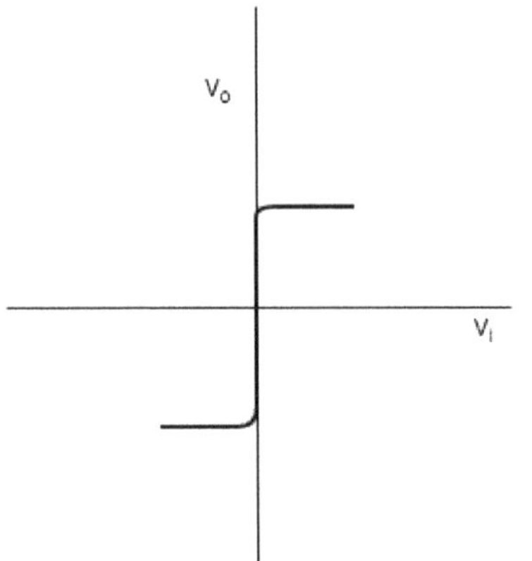

Open Loop gain of ideal Operational Amplifier is infinite.

Closed Loop Op-Amp: Input of such amplifier is largely in control of its output voltage. They are often called as feedback Operational Amplifiers. They are of two types:

1. **Negative Feedback:** When output voltage has opposite phase to that of input and subtracts a part from input voltage such amplifiers are negative feedback amplifiers. Negative Feedback Op-Amp has amplifying use.

2. **Positive Feedback:** Such Op-Amps are used in wave generators. Positive feedback is those where output voltage has an additive effect on input voltage or output voltage and input voltage are in phase to each other.

5.3 Open Loop Operational Amplifiers

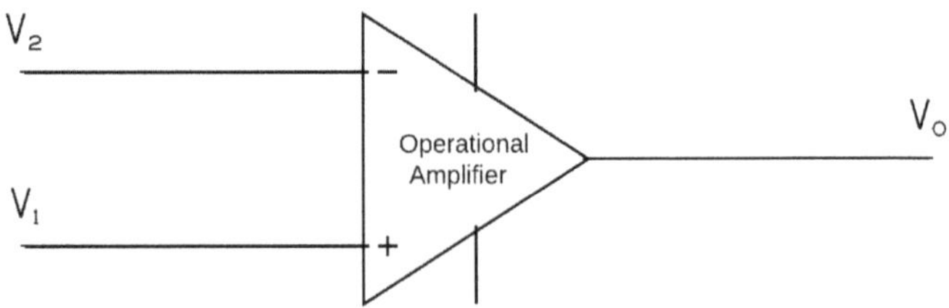

Take Operational Amplifier of Gain, output voltage as V_o. Input voltage is V_i.

Here $V_i = V_1 - V_2$. Where positive terminal of Op-Amp is called as non-inverting terminal and negative terminal of Op-Amp is called as inverting terminal.

As

$$A_{0L} = \frac{V_0}{V_i} = \frac{V_0}{V_1 - V_2}$$

So $V_0 = A_{0L}(V_1 - V_2)$

Open loop gain of Op-Amp is infinite

So V_0 is infinite.

$$V_0 = V_{saturation}$$

This is the saturation value of output voltage. The Highest Value of V_0 is $V_{saturation}$ in *figure 2* and *figure 3*.

5.4 Closed Loop Operational Amplifiers

In this chapter we will mainly focus on negative feedback closed loop amplifiers. Here concept of Inverting, Non-Inverting and Differential Operational Amplifiers arises. We are going to discuss them in detail and also see the notion of virtual short.

Concept of Virtual Short:

As we have seen that $A_{open\ loop}$ is infinite. We can say that V_{input} is zero. That means $V_1 - V_2 = 0$

$V_1 = V_2 = V_p$

This means that op-amp terminals are virtual ground. They are in same potential as gain is very high.

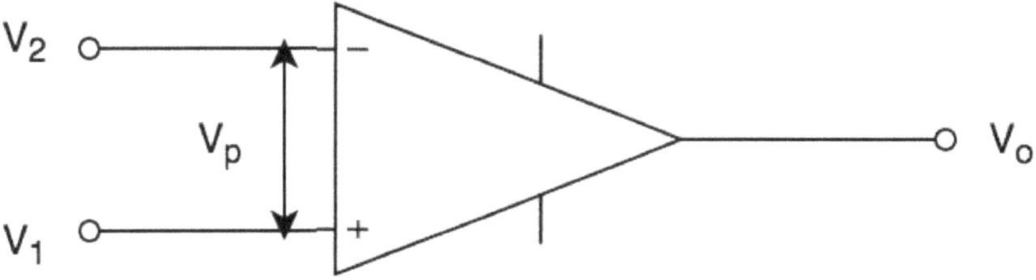

Fig. 5.1: Concept of virtual ground

Inverting Amplifiers:

When non-inverting terminal of an operational amplifier is grounded and input voltage (V_i) is applied across inverting terminal then this configuration is called as inverting operational amplifier. In closed loop it is shown in *figure 5*.

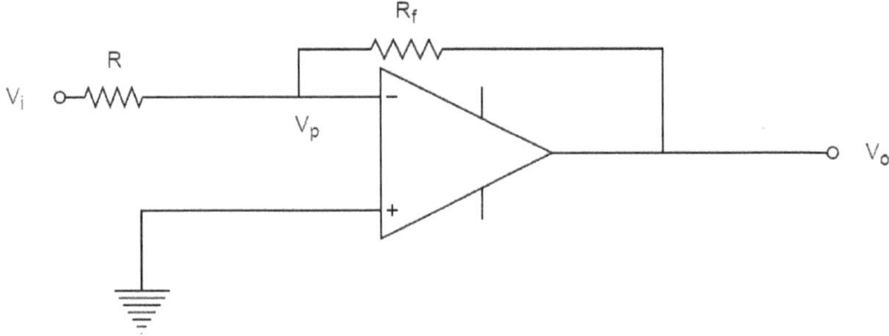

Fig. 5.2: Inverting Operational Amplifier

V_p Is virtual grounded point shown just to illustrate the fact that differential input across ideal Operational Amplifier is zero? In order to find the Gain of Operational Amplifier in inverting mode we proceed with KCL.

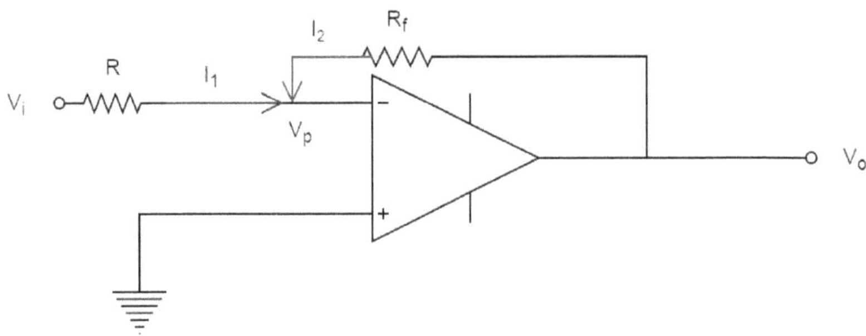

Fig. 5.3: Inverting Operational Amplifier

This means

$$I_1 = I_2$$

$$\frac{V_p - V_i}{R} = \frac{V_0 - V_p}{R_f}$$

Or

$$A = \frac{V_0}{V_i} = -\frac{R_f}{R}$$

So in inverting operational amplifier gain is always negative.

Non-inverting Operational Amplifier:

This type of amplifier suggests that input voltage is applied to non-inverting input and inverting input is however grounded. Non-inverting Op-Amp is shown in *figure 7*.

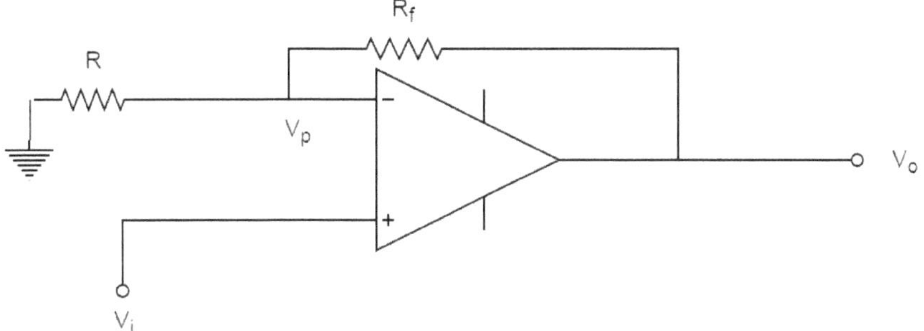

Fig. 5.4: Non-Inverting Operational Amplifier

Here too with KCL we can make

$$\frac{V_p - 0}{R} = \frac{V_0 - V_p}{R_f}$$

As here V_p is V_i so that differential input can be zero across Op-Amp. So

$$A = 1 + \frac{R_f}{R}$$

This is gain of Non-inverting Amplifier.

Differential Amplifier:

In Below figure we represent a differential amplifier. In this case input is given along both the terminals.

V_p Is equal to the voltage near non-inverting terminal of Op-Amp so that differential input is always zero.

$$V_p = \frac{V_2 R_2}{R_1 + R_2}$$

Now

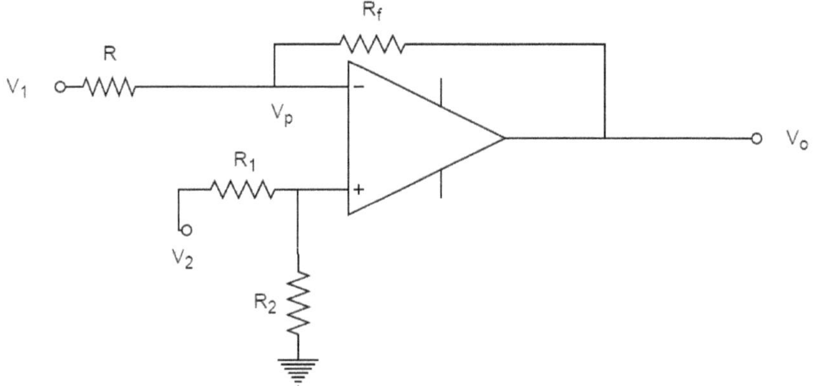

Fig. 5.5: Differential Op-Amp

If non-inverting input is zero then

$$V_{01} = -\frac{V_1 R_f}{R}$$

And if inverting input is zero then

$$V_{02} = \left(1 + \frac{R_f}{R}\right) V_p$$

$$V_{02} = \left(1 + \frac{R_f}{R}\right)\left(\frac{V_2 R_2}{R_1 + R_2}\right)$$

Total Input for $R = R_1$ and $R_f = R_2$

$$V_0 = V_{01} + V_{02}$$

$$V_0 = -\frac{R_f}{R}(V_2 - V_1) = -\frac{R_f V_i}{R}$$

Where V_i is differential input.

This shows that operational Amplifier can be used as subtractor with this configuration of resistors.

Generalized Differential Amplifier:

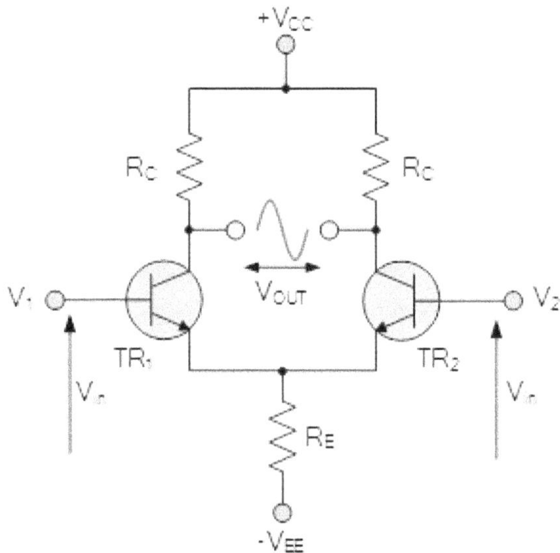

Fig. 5.6: Differential Operational Amplifier internal

This circuit represent the working of Differential Amplifier. Two transistors TR_1 and TR_2 have their collectors and emitters are connected to power supply V_{cc} and emitter supply $-V_{EE}$. Common input is fed to base of transistors. However from arrangement both input voltages are out of phase to each other.

V_{out} Is output taken along collectors of two transistors? Originally, if forward bias of one transistor is increased then forward bias of other will decrease. If input voltages are same then current will be constant across R_E.

If input voltages are same then difference of output voltages across two collectors will also same as current is always out of phase in two collector terminals. This is known as the *Common Mode of Operation* with the **common mode gain** of the amplifier being the output gain when input voltages are same.

Ideal Operational amplifiers can be regarded as having one output that ignore any change in common mode signals and resists any change to output signals with changing input signals but not their difference. This should mean that if an identical signal is somehow applied to both the inverting and non-inverting input terminals should not change the output.

However, in real amplifiers there is always some variation. For this we have another term, ratio of the change to the output voltage to the change in the common mode input voltage is called the **Common Mode Rejection Ratio** or **CMRR** in short form.

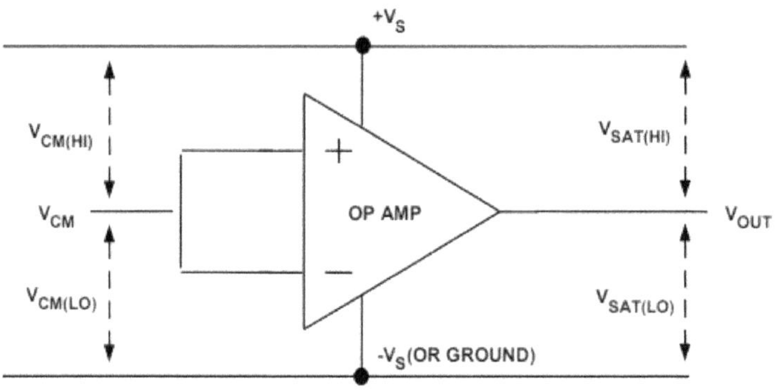

Fig. 5.7: Common mode operation

5.5 Parameters of Operational Amplifiers

Input Bias Current:

As operational amplifiers are made of BJTs or FETs the input terminals of operational amplifier are base or gate pins of transistors also shown in *figure* 8. Base current flows in transistor for its working along with collector and emitter currents. These currents across two terminals of operational amplifiers are called input bias currents.

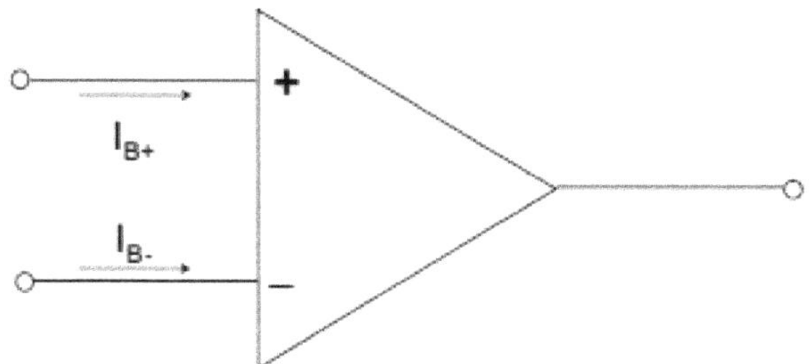

Fig. 5.8: Input Bias currents

In *figure* 11 I_{B+} and I_{B-} are input bias currents.

Input Offset current:

Input offset current is difference between input bias current.

$$I_{offset\ current} = I_{B+} - I_{B-}$$

Common Mode Rejection Ratio:

CMRR is the ability of op-amp to reject noise.

CMRR is ratio of differential voltage gain to common mode voltage gain.

$$CMRR = \frac{A_d}{A_{CM}}$$

Power Supply Rejection Ratio:

Input Offset voltage V_{io} changes with change in supply voltage V_{cc}. This is called Power Supply Rejection ratio.

$$PSRR = \frac{\Delta V_{io}}{\Delta V}$$

Its units are $\mu V/V$

Slew Rate:

Slew rate is maximum rate of change of output voltage per unit time. Its units are *V/μs*

Input Resistance:

It is resistance of IC chip measured from inverting or non-inverting terminal and output terminal is grounded. Ideal value of input resistance is infinite.

Input capacitance:

It is the capacitance measured from inverting or non-inverting terminal when output terminal is grounded.

Originally its value is $0pF$.

5.6 Ideal Op-Amp as Integrator

figure 12 shows how Op-Amp can be used as integrator

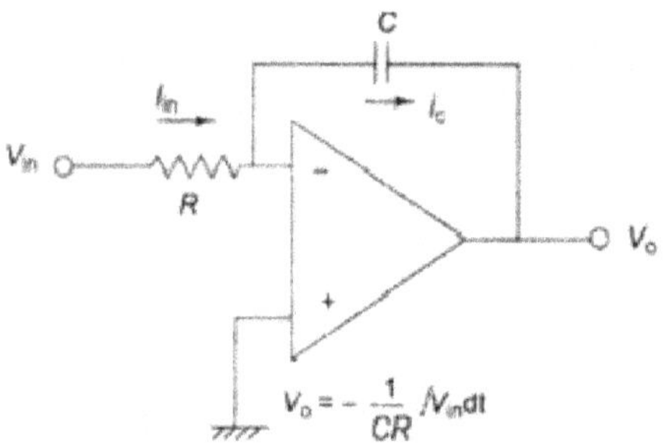

Figure 5.9: Op-Amp as integrator

$$I_{in} + I_c = 0$$

$$\frac{V_{in}}{R} = -C\frac{dV_o}{dt}$$

$$V_o = \frac{-1}{RC}\int_0^t V_{in}dt$$

5.7 Ideal Op-Amp as Differentiator

Op-Amp can be used as differentiator in this way

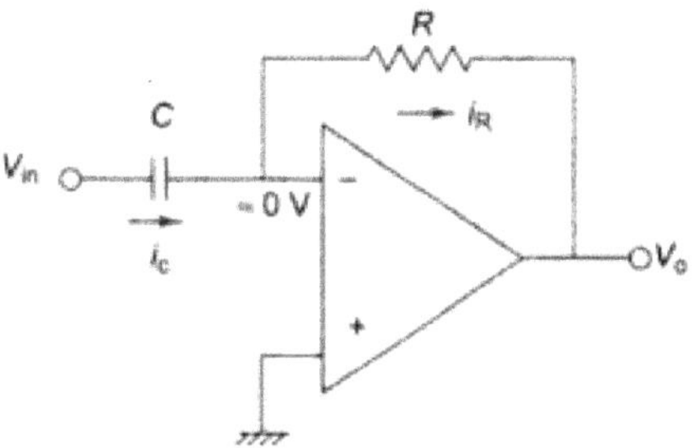

Figure 5.10: Op-Amp as differentiator

$$i_R = i_c$$

$$\frac{-V_o}{R} = C\frac{dV_{in}}{dt}$$

$$V_o = -RC\frac{dV_{in}}{dt}$$

Some Solved Problems

1. **What would input-output voltage characteristics will appear of an ideal Op-Amp with of 80V**

   $V_0 = V_{saturation}$ And characteristics of Op-Amp appear to be like

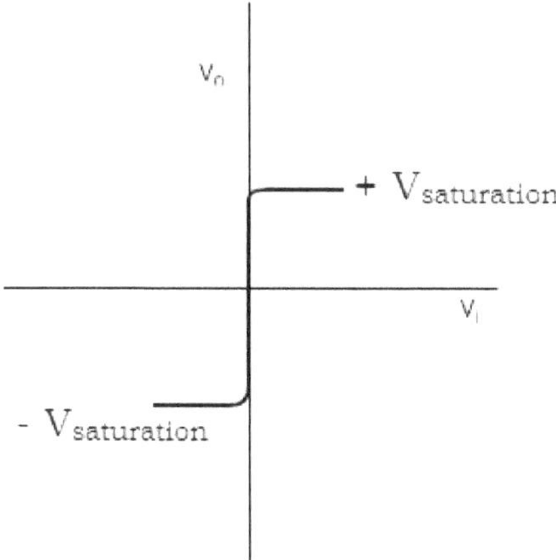

2. **In this circuit figure out current across resistor in A and B**

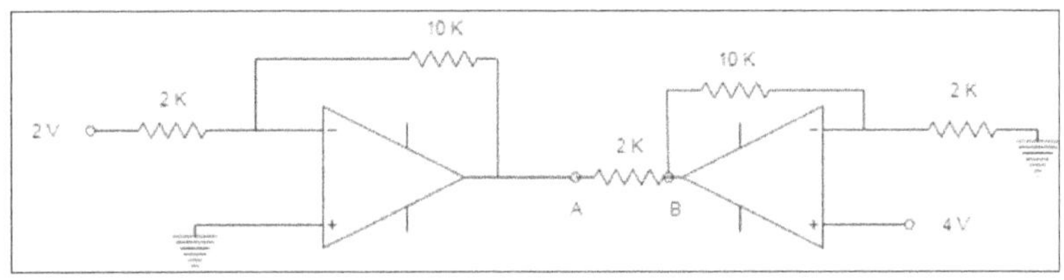

Across first Op-Amp output voltage is $V_{01} = -10V$

$V_{02} = 24V$

As gain in inverting and non-inverting amplifier is

$$A = \frac{V_0}{V_i} = -\frac{R_f}{R} \quad \text{And} \quad A = 1 + \frac{R_f}{R} \quad \text{and here } R_f \text{ in both cases is 10K}$$

and R in both cases is 2K.

so across resistor across A and B

current is $\dfrac{24 + 10}{2} = 17\,mA$

3. **Find the overall voltage across the final Operational Amplifier if input voltage is of form $V_i = (200mV) \sin(60t)$?**

Note that for non-inverting mode output voltage V_0 is

$$V_0 = \left[1 + \frac{R_f}{R_i}\right] V_i$$

And for inverting mode it is

$$V_o = \left[1 + \frac{R_f}{R_i}\right] V_i$$

So for first op-amp we get output voltage to be

$$V_{01} = 0.9V$$

And after 2nd Op-Amp it is

$$V_{02} = -3.5 \times V_{01} = -3.15\ V$$

And after 3rd Op-Amp it is

$$V_0 = -3.5 \times -3.15 = 11.025V$$

This will be the Output Voltage after 3rd Op-Amp

5.8 Applications of Operational Amplifiers

Op-Amp as Inverter

When non-inverting terminal of an operational amplifier is grounded and input voltage (V_i) is applied across inverting terminal then this configuration is called as inverting operational amplifier. In closed loop it is shown in *figure*.

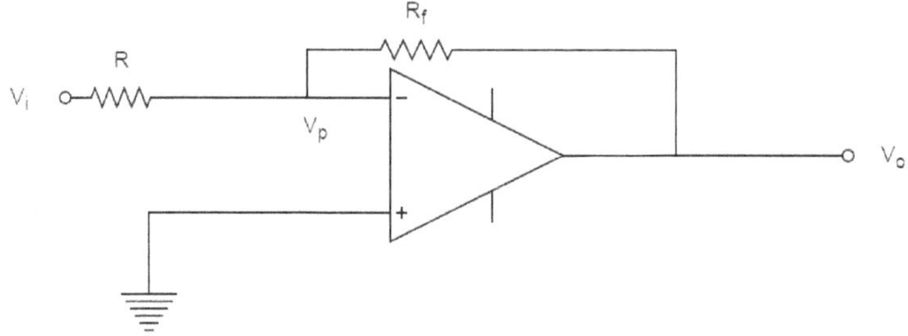

Figure 5.11: Inverting Operational Amplifier

V_p Is virtual grounded point shown just to illustrate the fact that differential input across ideal Operational Amplifier is zero? In order to find the Gain of Operational Amplifier in inverting mode we proceed with KCL.

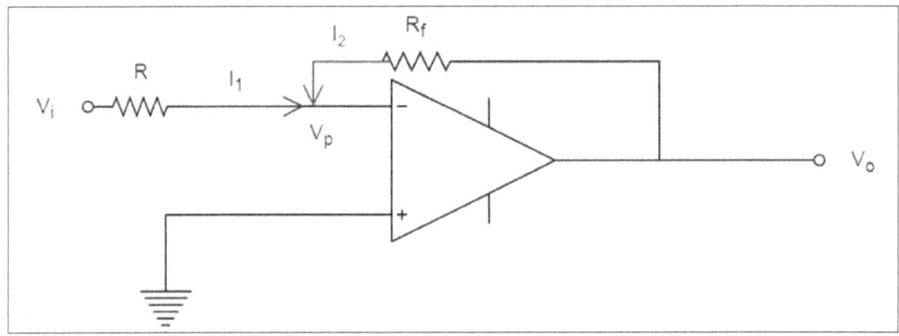

Figure 5.12: Inverting Operational Amplifier

This means

$$I_1 = I_2$$

$$\frac{V_p - V_i}{R} = \frac{V_0 - V_p}{R_f}$$

Or

$$A = \frac{V_0}{V_i} = -\frac{R_f}{R}$$

So in inverting operational amplifier gain is always negative.

Some applications of Inverting Amplifiers are:

- Inverting amplifier is use full for voltage adder or summing amplifier

- Inverting amplifier is applicable for the scaling summer amplifier.

- It is applicable for balanced amplifier.

Inverting Amplifier as Linear Current Output Device

Op-Amp also finds wide application as a current supplying device. This is accomplished by placing the load in the feedback loop as in figure.

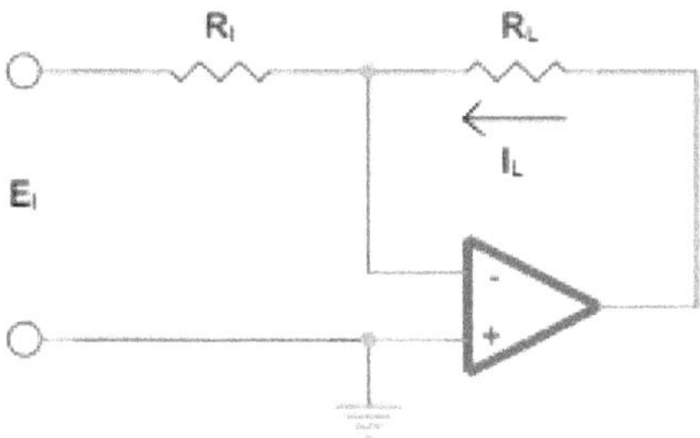

Fig. 5.13: Inverting Amplifier as Linear Current Output Device

Since the inverting input is ground potential, the current through R_I is $\dfrac{E_I}{R_I}$. No current flows into the inverting input, so $I_L = -\dfrac{E_I}{R_I}$. Which is independent of RL. In similar configurations, the inverting amplifier can serve as a linear meter amplifier or deflection coil driver. Input impedance is RI as before.

Inverting Amplifier as Comparator

V_p Is equal to the voltage near non-inverting terminal of Op-Amp so that differential input is always zero.

Now

$$V_p = \frac{V_2 R_2}{R_1 + R_2}$$

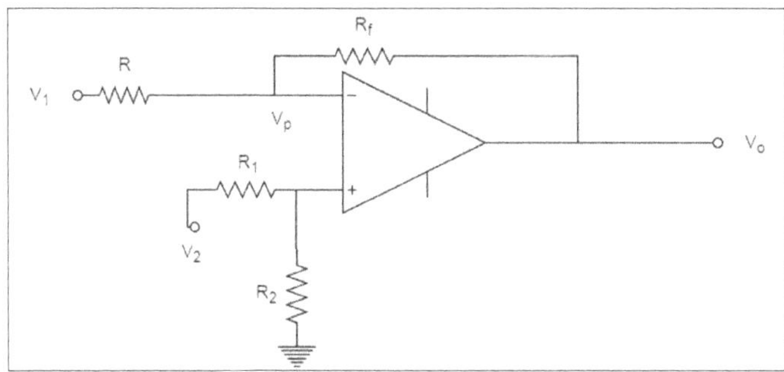

Fig. 5.14: Differential Op-Amp

If non-inverting input is zero then

$$V_{01} = -\frac{V_1 R_f}{R}$$

And if inverting input is zero then

$$V_{02} = \left(1 + \frac{R_f}{R}\right) V_p$$

$$V_{02} = \left(1 + \frac{R_f}{R}\right)\left(\frac{V_2 R_2}{R_1 + R_2}\right)$$

Total Input for $R = R_1$ and $R_f = R_2$

$$V_0 = V_{01} + V_{02}$$

$$V_0 = -\frac{R_f}{R}(V_2 - V_1) = -\frac{R_f V_i}{R}$$

Where V_i is differential input.

This shows that operational Amplifier can be used as subtractor with this configuration of resistors.

Non-inverting op-Amp

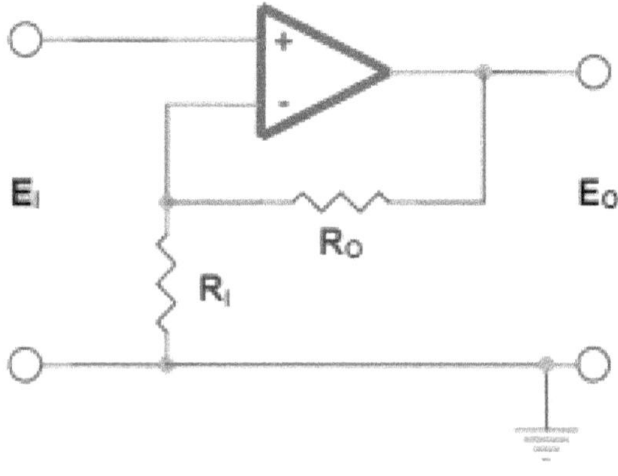

Fig. 5.15: Non-Inverting Op-Amp

This can be redrawn to show similarity of Inverting Operational amplifier with voltage follower.

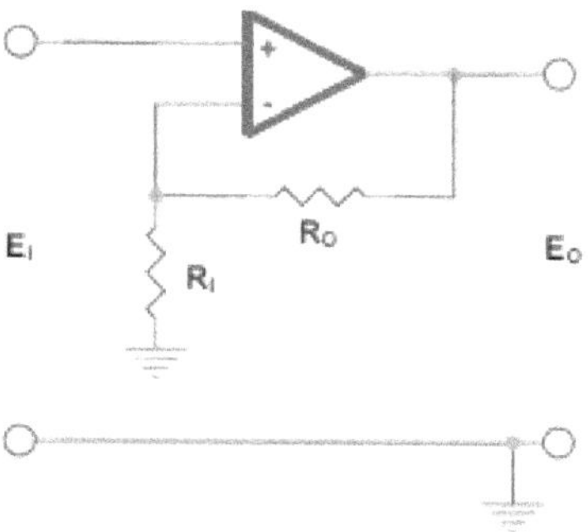

Fig. 5.16: Non-Inverting Op-Amp as Voltage Follower

As no current flows in R_1 and R_0 so this is a voltage divider circuit.

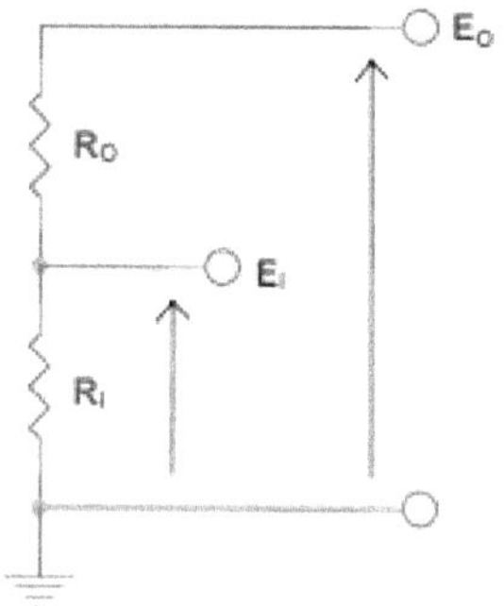

Fig. 5.17: Non-Inverting Op-Amp as Voltage Follower

The same voltage must appear at the inverting and non-inverting inputs, so that:

$$E_+ = E_- = E_I$$

Also

$$E_I = \frac{R_I}{R_I + R_o} E_O$$

$$\frac{E_O}{E_I} = \frac{R_I + R_o}{R_I} = 1 + \frac{R_o}{R_I}$$

Some applications of Inverting Op-Amp

- A non inverting amplifier uses a voltage divider bias negative feedback connection.

- Here the voltage gain is always greater than 1. So can be used in circuit as amplifier.

Op-Amp as Voltage Follower

A voltage follower is also sometimes referred as a unity gain amplifier or a buffer amplifier and also an isolation amplifier is an op-amp circuit which has a voltage gain of 1.

This means that the operational amplifier does not amplify the input signals. The reason it is called as a voltage follower is because the output voltage always follows the input voltage.

An op-amp circuit is very high input impedance device. This high input impedance is a reason voltage follower is used. The load resistance demands and draws a huge amount of current. This causes a huge amount of power to be drawn by the power sources. Voltage followers also referred as a voltage buffer.

Some applications of Voltage Follower

- High input resistance and a very low output impedance

- Voltage followers are mainly used to isolate stages from each other.

- Voltage follower is known as a voltage buffer.

Op-Amp as Integrator

Op-Amp can be used as integrator

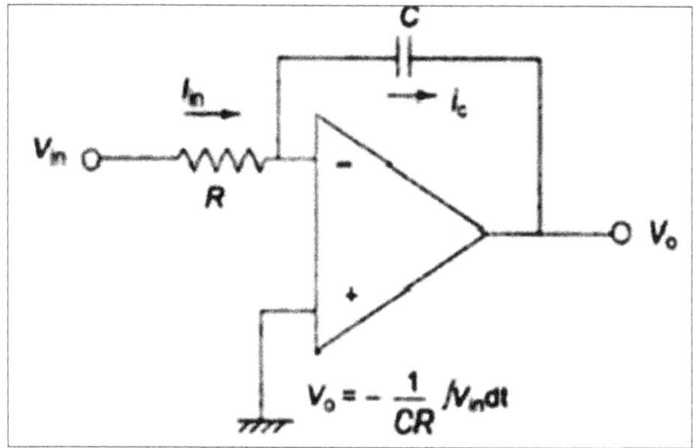

Fig. 5.18: Op-Amp as integrator

$$I_{in} + I_c = 0$$

$$\frac{V_{in}}{R} = -C\frac{dV_o}{dt}$$

$$V_o = \frac{-1}{RC}\int_0^t V_{in}\,dt$$

Op-Amp as Differentiator

Op-Amp can be used as differentiator in this way

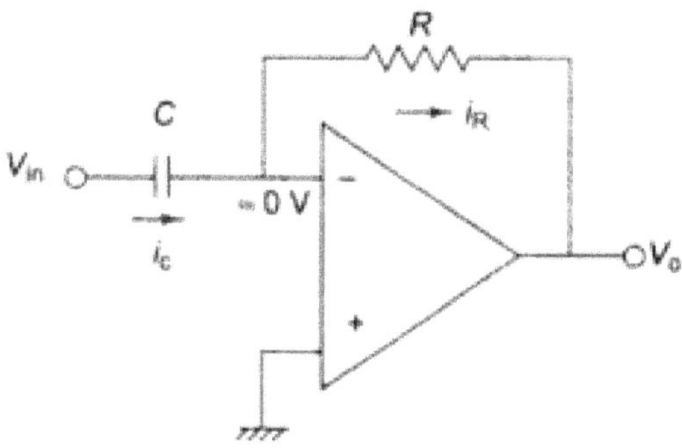

Fig. 5.19: Op-Amp as differentiator

$$i_R = i_c$$

$$\frac{-V_o}{R} = C\,\frac{dV_{in}}{dt}$$

$$V_o = -RC\,\frac{dV_{in}}{dt}$$

Voltage Adder

In a great many practical applications the input to the inverting amplifier is more than one voltage. The simplest form of multiple inputs is shown in figure.

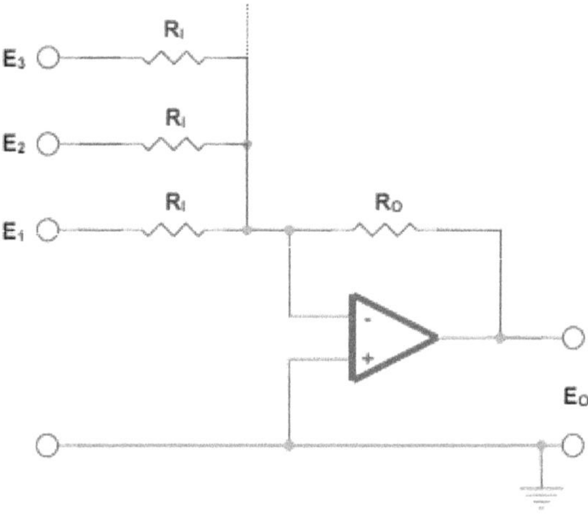

Current in the feedback loop is the algebraic sum of the current due to each input.

$$E_o = \frac{-R_o}{R_I}\,(E_1 + E_2 + E_3 + \cdots)$$

Scaling Summer Circuit

A more general form of adder allows scaling of each input before addition. Here Each input "sees" its respective input resistor as the input resistance.

$$E_o = -R_o\left(\frac{E_1}{R_1} + \frac{E_2}{R_2} + \frac{E_3}{R_3} + \cdots\right)$$

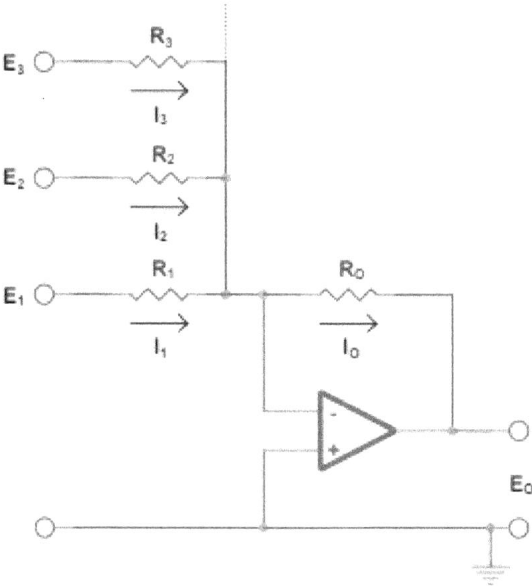

Some Applications of Summing Amplifier:

- Summing amplifier is known as a bipolar amplifier or a unipolar converter.

- Summing amplifier converts digital to analog converter.

SUMMARY

- Op Amp is a linear amplifying circuit commonly used in computers for mathematical operations like addition, subtraction and even integration and differentiation.

- Ideal Op-Amp has infinite gain that means $\dfrac{V_0}{V_i}$ is infinite.

- For real Op-Amp there is high voltage gain however not infinite and there is some value of input current. Output impedance too has a high value.

- Op-amp terminals are virtual ground. They are in same potential as gain is very high.

- When non-inverting terminal of an operational amplifier is grounded and input voltage (V_i) is applied across inverting terminal then this configuration is called as inverting operational amplifier

- In inverting operational amplifier gain is always negative.

$$A = \frac{V_0}{V_i} = - \frac{R_f}{R}$$

- Gain in non-inverting Amplifier is positive

- $A = 1 + \dfrac{R_f}{R}$ is gain in non-inverting Amplifier

- $V_0 = -\dfrac{R_f}{R}\left(V_2 - V_1\right) = -\dfrac{R_f V_i}{R}$ is gain in differential amplifier

- Op-Amp can be used as integrator, differentiator, summing and for comparing purposes.

- $V_o = \dfrac{-1}{RC}\displaystyle\int_0^t V_{in}\,dt$ is output voltage in case of integrator

- $V_o = -RC\,\dfrac{dV_{in}}{dt}$ is output voltage of op-amp when used as differentiator

SHORT ANSWER TYPE QUESTION

1. Determine V_0 if in open loop differential amplifier V_{in1} and V_{in2} are given as shown in figure and gain of the amplifier is 20,000

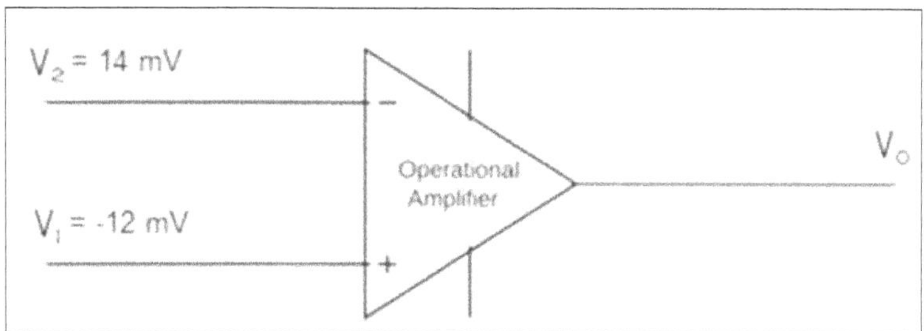

2. Find the maximum and minimum output voltage if all the feedback resistors ranges from 35 $k\Omega$ to 100 $k\Omega$ and R_1, R_2 and R_3 resistors ranges from 10 $k\Omega$ to 1 Ω. Input voltage is 10mV sin (30t).

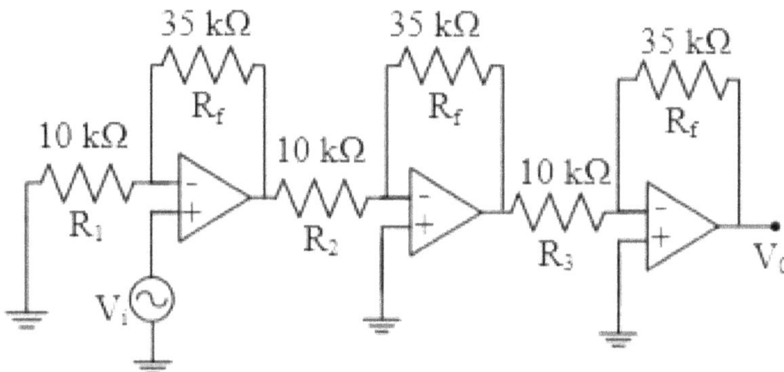

3. In the above question find the output voltage if input voltage is 100mV sin (30t). Also compare the result at time t=200ms and t=400ms

4. In the figure compare the results if the op-amp 1 is twisted so that non-inverting terminal becomes the inverting terminal and inverting terminal becomes non-inverting between the same things occurred to op-amp 3.

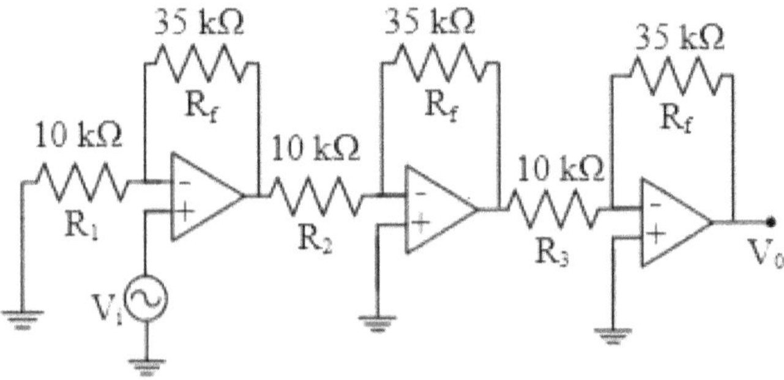

5. What could be the current in resistor R_2 and R_1 if output voltage is 44 *mV* and input voltage 1 is 11 *mV*.

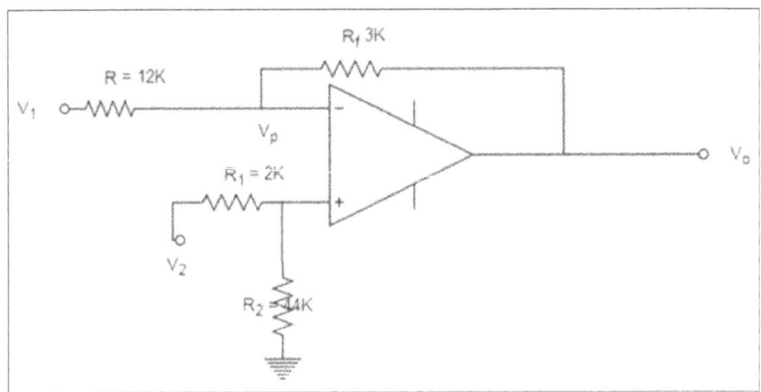

6. Again in the same configuration find the value of output voltage if input voltage 2 is given as 22mV and input voltage 1 is 2mV without the value.

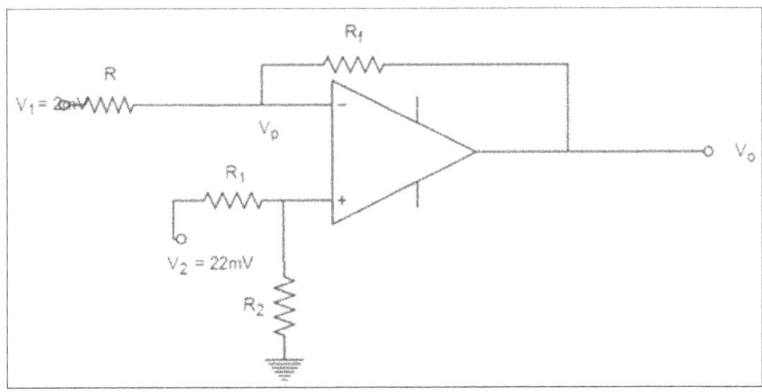

7. Find the value of I_c if I_{in} = 22mA and R = 44k. Output voltage is 23mV

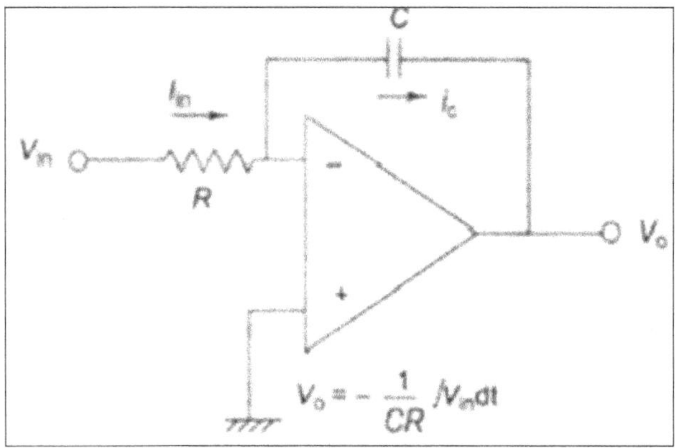

8. What could be the value of output voltage if relaxation time is 2ms and input voltage is of form 50cos(45t) + 10sin(30t) at time 3ms

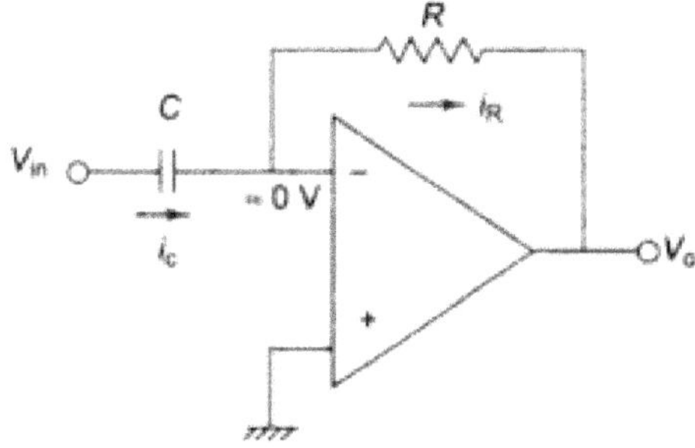